World Geomorphological Landscapes

Series editor

Piotr Migoń, Wrocław, Poland

For further volumes:
http://www.springer.com/series/10852

The Kuiseb River and Namib Sand Sea at Gobabeb

Andrew Goudie • Heather Viles

Landscapes and Landforms of Namibia

Springer

Andrew Goudie
School of Geography and the Environment
University of Oxford
Oxford
UK

Heather Viles
School of Geography and the Environment
University of Oxford
Oxford
UK

ISSN 2213-2090 ISSN 2213-2104 (electronic)
ISBN 978-94-017-8019-3 ISBN 978-94-017-8020-9 (eBook)
DOI 10.1007/978-94-017-8020-9
Springer Dordrecht Heidelberg New York London

Library of Congress Control Number: 2014945957

Printed on acid-free paper

Springer is part of Springer Science+Business Media (www.springer.com)

Series Editor Preface

Landforms and landscapes vary enormously across the Earth, from high mountains to endless plains. At a smaller scale, nature often surprises us creating shapes which look improbable. Many physical landscapes are so immensely beautiful that they received the highest possible recognition—they hold the status of World Heritage properties. Apart from often being immensely scenic, landscapes tell stories which not uncommonly can be traced back in time for tens of million years and include unique events. In addition, many landscapes owe their appearance and harmony not solely to the natural forces. For centuries, and even millennia, they have been shaped by humans who have modified hillslopes, river courses, and coastlines, and erected structures, which often blend with the natural landforms to form inseparable entities.

These landscapes are studied by geomorphology—'the Science of Scenery'—a part of Earth Sciences that focuses on landforms, their assemblages, surface and subsurface processes that moulded them in the past and that change them today. To show the importance of geomorphology in understanding the landscape, and to present the beauty and diversity of the geomorphological sceneries across the world, we have launched a book series *World Geomorphological Landscapes*. It aims to be a scientific library of monographs that present and explain physical landscapes, focusing on both representative and uniquely spectacular examples. Each book will contain details on geomorphology of a particular country or a geographically coherent region. This volume presents the geomorphology of Namibia—a country that hosts superb landforms, many being the best examples of their kind in the world. Endless sand seas, tall inselbergs, majestic river canyons, pans teeming with wildlife—they can all be found across Namibia. Since Namibia is relatively easy to navigate, the book is not only suitable for scientists and students of Geography and Earth Science, but can also provide guidance to holidaymaking geoscientists as to where to go to enjoy the very best scenery.

The World Geomorphological Landscapes series is produced under the scientific patronage of the International Association of Geomorphologists (IAG)—a society that brings together geomorphologists from all around the world. The IAG was established in 1989 and is an independent scientific association affiliated with the International Geographical Union (IGU) and the International Union of Geological Sciences (IUGS). Among its main aims are to promote geomorphology and to foster dissemination of geomorphological knowledge. I believe that this lavishly illustrated series, which sticks to the scientific rigour, is the most appropriate means to fulfill these aims and to serve the geoscientific community. To this end, my great thanks go to Professors Heather Viles and Andrew Goudie for adding this book to their agendas and delivering such an exciting illustrated story to read and admire. The thanks are more than customary. Many years ago, I was invited to join Heather and Andrew on one of their Namibian research trips and benefited most from their expert knowledge of the country, now shared with the global geomorphological community.

Piotr Migoń

Acknowledgments

Over the years, we have been grateful to a number of people for their assistance and valued company in the field in Namibia, including Dick Grove, the late Charles Koch, Abi Stone, Dave Thomas, Jennifer Lalley, Frank Eckardt, Alexander Shaw, Brett Smith, Piotr Migoń, Susan Conway, Mary Bourke, Mark Taylor, Ian Livingstone, Mary Seely, Stuart Neumann, Alice Goudie, Amy Beasley, Joh Henschel, and Andy Watson. We have particularly valued the hospitality that the Gobabeb research facility has offered to ourselves and our students over many years. We are also grateful to some of our German colleagues, including the late H. Besler, K. Heine, W.-D. Blümel, M.W. Buch, and B. Eitel for so kindly remembering to send us offprints relating to their own valued contributions to our understanding of the geomorphology of Namibia. We are grateful to Elsevier for permission to reproduce figures 1.5, 3.4, 8.1, 8.2, 8.3, 10.5, 10.6, 18.2, 18.1, 18.6, 20.3, 20.4, and 26.3; to Wiley for figures 1.6, 2.5, 2.6, 8.6, 10.2, and 11.3; and to the National Archives of Namibia for figure 1.9.

Contents

Part I

Introduction to Namibia and its Landscapes

1 Landscapes

Abstract

Namibia has a wide range of landscapes and the classification provided in the *Atlas of Namibia* is adopted by way of introduction to their nature and diversity. This is followed by a description of three main landscape regions: the Namib Desert, the Kalahari Desert, and the Great Escarpment. Finally, the chapter includes a regional analysis of the river systems and the coastline of Namibia, with feature boxes on two iconic landscapes, the Fish River Canyon and Sandwich Harbour.

1.1 Landscape Types

Namibia is a vast and varied country with wonderful landscapes and landforms, many of which have been engagingly portrayed in words and pictures by Swart and Marais (2009). It is particularly notable because of the richness and beauty of its desert landforms, and because of what it can tell us about the long-term tectonic history and climate of this part of Africa. The major controls on landscape evolution are tectonics (and its influence on geology) and climate (and its influence on ecology) and their dynamic interrelationships over a range of timescales. However, before introducing the outlines of tectonic and geological histories (Chap. 2) and the dynamics of climate and ecosystems (Chaps. 3 and 4) this chapter introduces the major characteristics of Namibian landscapes and their diversity, including the nature of the two great deserts (the Namib and the Kalahari), the Great Escarpment which runs down its spine, its rivers, and its long coastline.

In the second part of the book we present a series of regional studies illustrating some of the most dramatic and interesting landforms and landscapes of the country, and these are approximately arranged from north to south (Fig. 1.1). Chapter 25 is not marked on Fig. 1.1, as the phenomena it describes occur over much of Namibia. We have chosen to include landscapes and landforms for which there is a good array of scientific literature and which reflect the diversity of landscape types in Namibia. Other important and much visited landscape features, such as the Fish River Canyon and Sandwich Harbour, are described in boxes in Chap. 1, while another characteristic landform type—sandstone-capped mesas—is featured in box 3 in Chap. 2.

Covering an area of about 823,680 km^2, Namibia is up to 1,320 km long, and 1,440 km wide. It is, however, sparsely populated with only around 2 million inhabitants. To the north it is bounded by Angola and Zambia, to the east by Botswana, to the south by South Africa, and to the west by the cold waters of the Atlantic Ocean. Much of Namibia consists of a wide plateau at 900–1,300 m above sea level (Fig. 1.1). This plateau is bounded on the west by a large escarpment and on the east by the Kalahari Basin. Bordering the Atlantic in the west is the lower-lying coastal plain of the hyper-arid Namib Desert (Van Zyl 1992). Van der Merwe (1983) estimated that plains were the dominant landscape of Namibia, covering over 45 % of the country, with mountains covering c 19 %, dunes just under 14 %, plains with scattered hills over 13 %, and hills just under 8 %. Wellington (1967) divided the Namibian landscape into three main types—the Namib Desert, the Plateau Hardveld and the Kalahari Sandveld, within which he identified a number of more specific landscape types.

Recently, *The Atlas of Namibia* (Mendelsohn et al. 2002) has identified a range of landscape regions in the country, providing a useful framework to describe the geomorphological diversity (Fig. 1.2). These regions are described here heading from north to south, apart from two areas (the Kalahari sandveld and the Great Escarpment) which span large distances from north to south in the eastern and western parts of the country respectively, with which we begin.

A. Goudie and H. Viles, *Landscapes and Landforms of Namibia*, World Geomorphological Landscapes,
DOI 10.1007/978-94-017-8020-9_1, © Springer Science+Business Media Dordrecht 2015

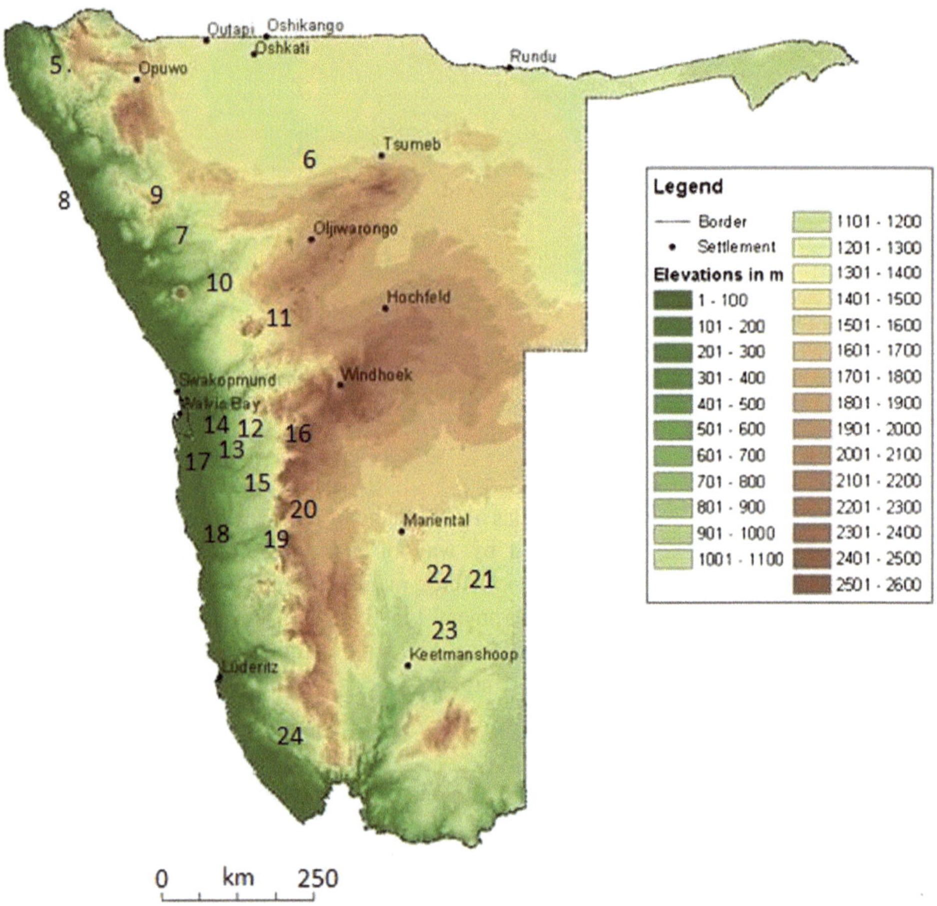

Fig. 1.1 The approximate locations of Chaps. 5–24 and the relief of Namibia (from Mendelsohn et al. 2002, p. 39 in http://www.uni.koeln.de/sfb389/e/e1/download/atlas_namibia/) (accessed 30th January 2014)

The *Kalahari sandveld* occupies a huge part of northern and eastern Namibia. It is a generally monotonous, flat, basin of sedimentation, much of which is characterised by aeolian landforms, including linear dunes and pans (Thomas and Shaw 1991). It is discussed further in Chap. 21. The *Escarpment* (or Great Escarpment), discussed later in Chap. 1, runs roughly parallel to the coast and divides much of the country up into two general landscapes: the low-lying coastal plain to the west, and the higher inland plateau to the interior. It is not a continuous feature, and is largely absent from the Central-Western Plains.

In the far north east is a small area called the *Caprivi Floodplains*, created by the Zambezi and Kwando rivers and consisting of a network of channels, spectacular oxbow lakes and grasslands. In the late Pleistocene it may have been occupied by a lake, called Lake Caprivi (Shaw and Thomas 1988). Further east, the *Okavango Valley* occurs as a narrow strip along Namibia's northern border. The *Karstveld* of northern Namibia covers a scatter of areas in the east and west and is underlain by soluble carbonate rocks, including limestones and dolomites, and has an array of karstic forms including caves and sinkholes, of which Guinas and Otjikoto are the most dramatic examples (see Chap. 6). *Pans* are a typical Namibian landscape element, represented most notably in the north by Etosha Pan (see Chap. 6). The *Cuvelai system* which lies between Etosha Pan and the Angolan border, is dominated by a network of curious, shallow channels, called 'Oshanas', which in wet years obtain much of their water from the Angolan Highlands. The *Kunene Hills* in the far north west of Namibia, sometimes called the Kaoko Highlands, are a rugged area of dissected ancient rocks, commonly 1,000–1,900 m above sea level (Sander 2002). The hills include the Baynes, Steilrand, and Zebra Mountains. The Zebra Mountains are so called because they consist of a mass of interlayered, relatively unaltered dark leucotroctolite with relatively altered, "white," anorthosite (Maier et al. 2013). Glacial features, originating in the Dwyka phase (c 200–270 million years ago), have been exhumed and are widespread. These include U-shaped valleys with striated walls. The *Etendeka Plateau* (discussed further in Chap. 9) consists of flat-topped hills underlain by great expanses of volcanic lava and some sedimentary Karoo age rocks. The lavas were spewed out when Africa and South America split apart some 132 million years ago (as discussed in more detail in Chap. 2). The *Kamanjab Plateau*, mostly underlain by ancient granites and gneisses, is in the north west of the country and is drained and dissected by the Huab and Ombonde rivers.

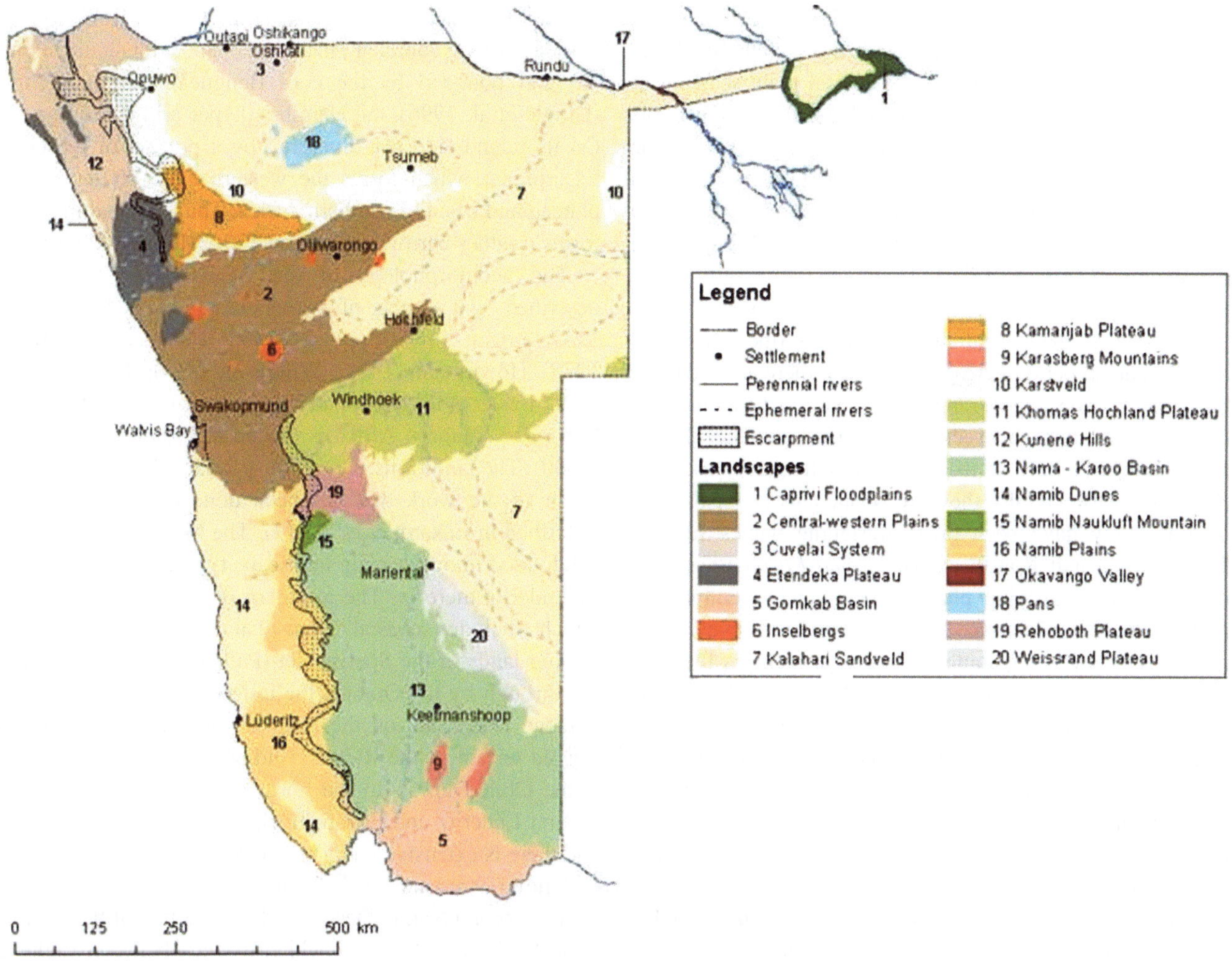

Fig. 1.2 The landscape divisions of Namibia (from Mendelsohn et al. 2002, p. 14, in http://www.uni-koeln.de/sfb389/e/e1/download/atlas_namibia/) (accessed 30th January 2014)

A large landscape unit within central Namibia is the *Central-Western Plains*, much of which lies between 500 and 1,000 m above sea level, stretches inland from the Atlantic coast and has been formed by rivers such as the Khan, Omaruru, Swakop and Ugab cutting back eastwards into higher ground. The area is studded with upstanding granite hills called inselbergs (Mabbutt 1952). These *Inselbergs* are large, free-standing mountain masses that punctuate the Central-Western Plains, and include Brandberg (at 2,579 m, the highest point in Namibia), Erongo (see Chap. 11), Paresis and Spitzkoppe (see Chap. 10). The *Khomas Hochland Plateau* is located in the centre of the country around Windhoek, and consists of a ridge of rolling hills and deep valleys. Much of it lies at altitudes between 1,700 and 2,000 m above sea level and it receives sufficient rainfall to feed such rivers as the Nossob, the Kuiseb and the Swakop. It is the remnant of a once-great mountain chain created towards the end of the Damara stage (c 550 million years ago) as a result of the collision of continents. This area contains one somewhat anomalous landform curiosity, a field of late Pleistocene dunes neither connected with the Namib or the Kalahari, up to c 7 m tall, located at Teufelsbach, some 20 km south of Okahandja (Eitel et al. 2004). The *Naukluft Mountains*, which largely consist of limestones and shales, occur on the edge of the Great Escarpment within central Namibia. The mountains are highly dissected by small, steep valleys in which extensive spreads of calcareous tufa occur (see Chap. 20). The *Rehoboth Plateau* lies in the centre of the country to the south of Windhoek at an altitude of between 1,500 and 1,700 m above sea level. It is an area of inselbergs and rolling terrain underlain primarily by granites and complexes of metamorphic rocks.

The western parts of central and southern Namibia are dominated by sand and plains. The *Namib Sand Sea* (see Chap. 18) stretches for 400 km north from Lüderitz (now officially known as #Naminus) to Walvis Bay and is up to 100–140 km in width. It contains a wide array of large and mobile dunes. The *Namib Plains* (see Chaps. 12 and 14) consist of gravel and gypsum covered surfaces, rocky outcrops and hills, which together with the Namib Sand Sea make up a large proportion of the coastal plain seaward of the escarpment. Sand ramps are often banked up against hills (Bertram 2003). The *Karas Mountains* of southern Namibia consist of uplifted blocks of sandstones, limestones and shales that rise up above the surrounding plains. The highest peak in the Gross Karas Mountains reaches 2,203 m above sea level. The *Gamchab Basin* is an area to the north of the Orange River, with large valleys created by river erosion. Over much of the area drainage densities (the amount of stream channel per unit area) are high and there are extensive fan systems. The *Islands*, of which there are 12 main ones, occur just offshore between Walvis Bay and the Orange River and have been noted for their rich guano resources (Watson 1930). Whilst they are small and inconspicuous features, they have been given intriguing names such as Plumpudding and Roastbeef. The *Nama-Karoo Basin* is a predominantly flat-lying plateau underlain by sedimentary rocks, which slopes from 1,400 m above sea level in the north to 900 m in the south. This region includes the Schwarzrand to the south of Maltahöhe. It is drained by rivers such as the Fish, which flows to the Orange. Some ancient inselbergs have been exhumed from beneath the former Late Proterozoic to Cambrian Nama sedimentary cover (Stengel 2000; Stengel and Busche 2002) either because of river erosion or groundwater-related weathering effects (Twidale and Maud 2013). Fossil landslides have been extensively developed (Stengel 2001), probably as a result of higher precipitation amounts than today. Brukkaros forms the only major mountain in this area (see Chap. 24). Finally, the *Weissrand Plateau* is an intriguing area of solution hollows (dayas), calcrete, aligned drainage and old dunes sandwiched between the Nama-Karoo Basin and the Kalahari sandveld (see Chap. 22).

1.2 The Namib and the Kalahari Deserts

Of the composite landscape types of Namibia, two of the largest and most important are the two great deserts, the Namib in the west and the Kalahari in the east. The Namib Desert landscape comprises a range of landscape types from hills to gravel plains and dune fields, whereas the Kalahari Desert is dominated by stabilised dunes (Fig. 1.3).

1.2.1 The Namib Desert

The Namib, one of the world's driest and most beautiful deserts, extends for more than 2,000 km and eighteen degrees of latitude along the Atlantic coast of southern Africa from the Olifants River in South Africa (latitude 32°S) to the Carunjamba River (latitude 14°S) in Angola. Being on the west side of the continent, in a zone of subsiding anticyclonic air, and bounded by the cool Benguela current offshore (Dingle et al. 1996), the Namib is hyper-arid (see Chap. 3). On its inland side it is bordered by a portion of the Great Escarpment which forms the western edge of the interior plateau and basin of southern Africa. Thus the Namib Desert forms a rather narrow strip some 120–200 km wide.

The geomorphology of the Namib Desert has been described by a number of workers (e.g. Gevers 1936; Cloos 1937; Logan 1960; Spreitzer 1965; Beaudet and Michel 1978; Hövermann 1978; Wilkinson 1990; Lageat 1994, 2000; Besler et al. 1994) and its context within the Cenozoic history of southern Africa is treated in Partridge and Maud (2000). The landforms in proximity to Gobabeb, the base for much of the work that has been done on the desert, are described in Eckardt et al. (2013). The Namib Desert can be subdivided into four main landscape types. In the area south of Lüderitz there is 'The Southern or Transitional Namib', which includes coastal Namaqualand and the diamond mining lands of the Sperrgebiet (Pallett 1995). This zone is cut through by the Orange, the last perennial river until the Kunene is reached on the Angolan border. It includes the rugged terrain of the Richtersveld and some areas of dunes —the Obib Dunes—to the north of the Orange. The area around Lüderitz and Elizabeth Bay has high velocity winds and there is extensive yardang development, rock fluting and deflation (Lancaster 1984; Corbett 1993).

The second Namib Desert landscape is that of the 'Namib Sand Sea' (see Chap. 18) which extends between Lüderitz and Walvis Bay and contains some of the world's biggest dunes. This area has been the subject of a detailed review by Lancaster (1989). The third Namib Desert landscape is the 'Central Namib Plains'. These lie between the Kuiseb River and more dissected terrain that lies to the north of the Brandberg. The plains have a low gradient of only 1° between the coast and the 1,000 m contour, and are studded with marble and dolerite ridges, some isolated inselbergs and complexes of shallow pans (Eckardt et al. 2001; Eckardt and Drake 2011). The plains show many windstreaks oriented with the easterly 'berg' winds which generate some dust plumes, which head out across the South Atlantic. Although the area is hyper-arid, the plains are also crossed by a very dense and intricate network of shallow drainage lines. This is very evident, for example, on the gently sloping rock surface immediately behind the coastal dunes between Walvis Bay and Swakopmund. Locally, ephemeral rivers such as the Swakop are more deeply incised into the plains, producing gorges and areas of badlands (called *gramadullas* or a moon landscape) (Fig. 1.3). In places these gorges truncate groundwater aquifers so that seepage occurs. This produces tufas made either of lime or of halite. The fourth landscape —'The Northern Namib and Skeleton Coast'—includes a

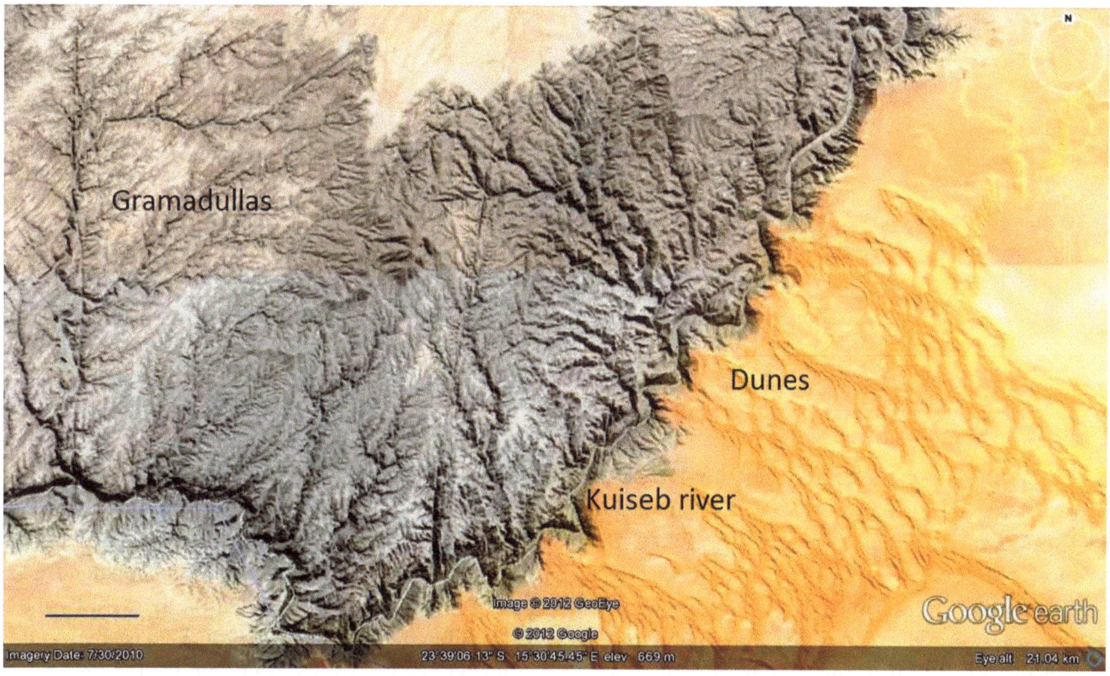

Fig. 1.3 Google Earth image of gramadullas to the north of the Kuiseb River. Scale bar 2 km (© 2013 GeoEye, Google)

dissected area of sandstone and lava hills, known as the Kunene Hills (also called Kaoko Highlands or Kaokoveld), together with some coastal dunefields (Lancaster 1982). The Kunene Sand Sea, which extends into Angola, is cut through by the perennial Kunene River (see Chap. 5).

It is also possible to divide the Namib Desert on the basis of its climate. Besler (1972), for instance, related weathering and other phenomena to the fog environment, introducing three divisions: the cool fog desert at the coast, the desert steppe in the east and the warm (alternate) fog desert in the middle.

One important landscape type, related to desert conditions, is found in between the Namib and Kalahari deserts in northern Namibia. Here, there are some quite extensive deposits of loess. Loess is a largely non-stratified and non-consolidated silt, containing some clay, sand, and calcium carbonate. It consists chiefly of quartz, feldspar, mica, clay minerals and carbonate grains in varying proportions. The grain size distribution of typical loess shows a pronounced mode in the range 20–40 μm and is generally positively skewed towards the finer sizes. It was the great German geographer, Ferdinand von Richthofen (1882, pp. 297–298), who had travelled to the classic deposits in China, who cogently argued that these intriguing deposits probably had an aeolian origin and that they were produced by dust storms transporting silts from deserts and depositing them on desert margins. Thus, it is likely that the Namibian loess deposits on the margins of the Kalahari and Namib Deserts have been produced from dust originating in these arid environments.

Namibian loess locations include the Opuwo basin and Omungunda in the Kaokoland area, where they were originally thought to be of late Holocene age (Brunotte and Sander 2000). However, Brunotte et al. (2009) have recently asserted that in the Opuwo area loess deposition commenced around 55,000 years ago (i.e. in the Pleistocene rather than the Holocene). Loess, up to 5 m in thickness, also forms a fill in large basins in the valleys of the Huab and Hoanib rivers in the Khorixas district (Eitel et al. 2001), and appears to be of largely late Pleistocene age. It is believed that the loess is formed from material transported by westward moving dust storms from the eastern Kalahari under drier conditions than today. Even today, dust is generated in substantial quantities from the surfaces of the Mkagadikgadi depression in Botswana (Washington et al. 2003), and the Etosha Pan in Namibia (Bryant 2003). The loess is now being eroded by water to give areas of badlands.

1.2.2 The Kalahari Desert

In the interior of southern Africa, much of it in Botswana but a substantial part in eastern Namibia, lies the Kalahari Desert (Thomas and Shaw 1991). This area was the subject of a major study by the Prussian geographer Passarge (1904), though most of his observations took place in what in his day was called Bechuanaland. However, he did describe the stratigraphy and landforms of the Gobabis area in eastern Namibia.

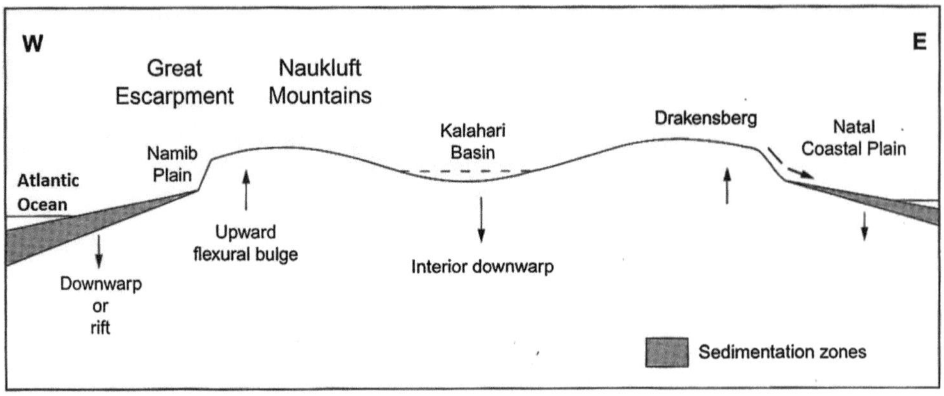

Fig. 1.4 The structural context of the Kalahari

It is difficult to say what precisely the borders of the Kalahari Desert are, not least because it has expanded and contracted during the last few million years. Much of it is a relict of a more extensive desert that once extended equatorwards well into the Congo Basin. It also merges with the Namib in the west and the Karoo in the south, and its boundary with the latter is often taken as the Orange River. The Kalahari, most of which lies at an altitude of around 1,000 m, derives its name from the Setswana word 'Kgalagale', which means 'always dry', but there are in a sense three Kalaharis, some drier than others:

(a) *The Kalahari dune desert in the arid south west interior of Botswana and adjoining parts of Namibia and South Africa.* The primarily summer rainfall is less than 200 mm per annum and is just sufficient to stabilize the plinths of a major field of dominantly linear dunes. The dune crests are often active.

(b) *The Kalahari region (or thirstland) approximately delineated in the north by the Okavango Swamps and in the south by the Limpopo and the Orange rivers.* This is an area of little or no surface drainage despite a relatively higher rainfall (c 600 mm per annum). Rates of groundwater recharge are very low (De Vries et al. 2000). It is almost entirely covered with grass and woodland, and has extraordinarily low relief.

(c) *The Mega-Kalahari, which is an extensive area consisting of a basin filled by the continental sediments of the Kalahari Beds.* This extends from the Orange River as far as the Congo. Precipitation may be as high as 1,500 mm, but it displays the evidence of former aridity in terms of the development of ancient dune systems, drainage alignment, and pans (Shaw and Goudie 2002).

The Kalahari contrasts with the Namib Desert because of its relatively high rainfall and because of its basinal form. Because the climate of the Kalahari is semi-arid to subhumid, most of it is not a true desert but an extensively wooded 'thirstland'. Over enormous distances the relief is highly subdued and the landscape monotonous. The Kalahari owes its gross form and subdued morphology to the fact that following the break-up of Gondwanaland it became an area of down-warping bounded on the west by the highlands of Namibia and Angola, and on the east by mountains such as the Drakensberg and Lubombo (Haddon and McCarthy 2005) (Fig. 1.4). It became a basin of sedimentation and this largely accounts for its flatness. The Kalahari Beds that fill this basin are often over 100 m in thickness and in parts of the Etosha region of northern Namibia they are over 300 m thick (Fig. 1.5). Differences in thickness are related to graben (fault) structures. The sediments consist of terrestrial conglomerates, breccias, clays, dune sands, diatomaceous interdune deposits, alluvium, calcretes, silcretes and marls (Wanke and Wanke 2007). Depositional settings included braided rivers, sheet flood areas, shallow lakes and pans, and dune systems.

Apart from its relict linear dunes (Thomas 1984) (see Chap. 21), the Kalahari contains large numbers of pans and associated leeward lunette dunes (Goudie and Thomas 1985) (see Chap. 23), together with two large closed depressions—the Etosha Pan of Namibia (see Chap. 6) and the Mkgadikgadi Depression of Botswana. Today these are major sources of dust plumes (Washington et al. 2003). In the mid to late Tertiary, palaeolake Etosha received water via the Cubango, Kunene and Cuvelai drainage systems. It largely dried up at about 4 million years (Ma) under conditions of progressively increasing aridity, though it still occasionally floods (Miller et al. 2010). The Kalahari shows excellent development of calcrete, silcrete and combinations of the two (Watts 1980; Nash et al. 1994; Shaw and Goudie 2004). The reason why calcretes in particular are so well developed probably relates to the structural context. The long history of gentle sedimentation within the Kalahari basin has created suitable conditions for the preservation of

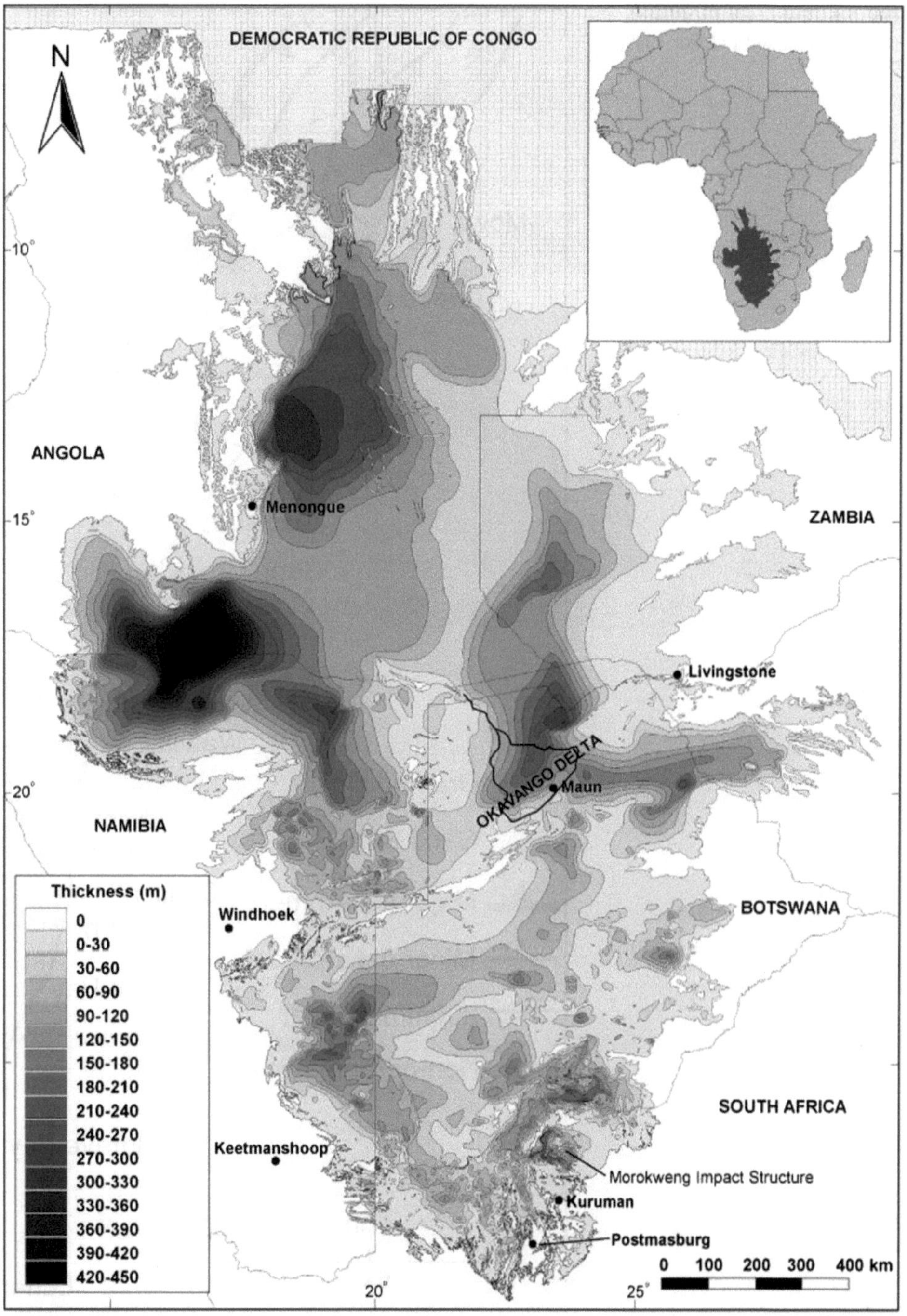

Fig. 1.5 The extent and thickness of the Kalahari Beds (from Haddon and McCarthy 2005, Fig. 1)

calcrete sequences, and the presence of ancient limestones and dolomites on the basin margins has supplied the necessary solutes.

1.3 The Great Escarpment

A third great landscape unit in Namibia is the Great Escarpment (Fig. 1.6), which lies inland of the Namib plains and rises up above them. This is one of the most important and conspicuous landforms of southern Africa (Kempf 2010). Such great escarpments 'dominate the landscape of many rifted margins and are among the largest topographic features on Earth and compare with orogenic belts in majesty, although they are associated with plate divergence rather than subduction' (Garzanti et al. 2014, p. 17).

The Namibian Great Escarpment is part of a feature that stretches all around Southern Africa from Mozambique in the east to southern Angola in the west. In Namibia it comprises ranges such as the Naukluft Mountains (see Chap. 20) and the Gamsberg and contains some deep gorges. Of these the most impressive is the Fish River Canyon

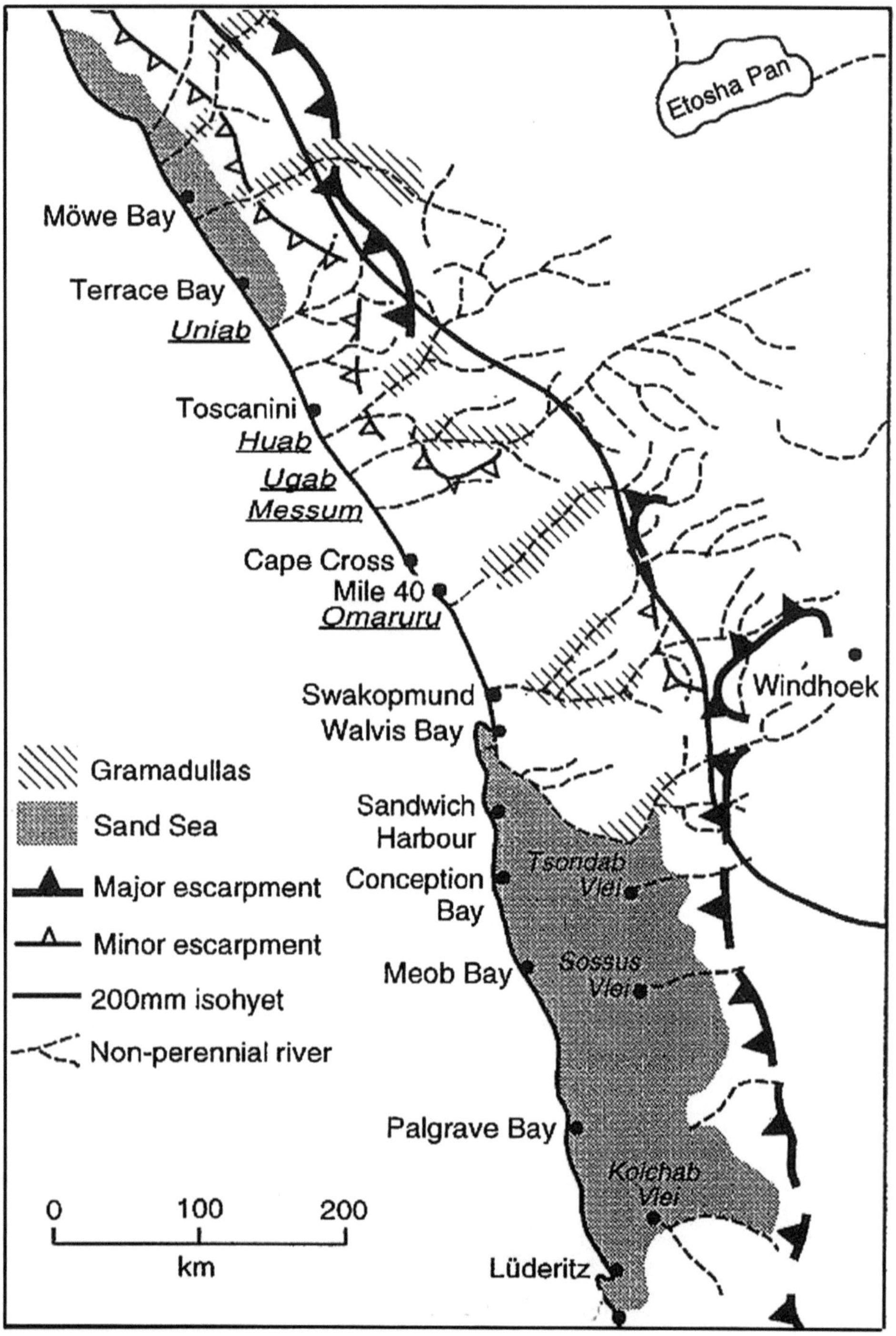

Fig. 1.6 The Great Escarpment (from Goudie and Eckardt 1999, Fig. 1)

(Mvondo et al. 2011). This is one of the largest canyons in the world, being well over 500 m deep and extending for over 50 km. It is partly an ancient graben feature but is also the result of uplift and incision following the breakup of Gondwanaland (Grünert 2000).

The flat-topped Gamsberg is part of the Great Escarpment that separates the Khomas Hochland to the east from the low-lying Namib plains to the west. With an elevation of 2,347 m above sea level it towers above the Khomas Hochland by 450 m and above the Namib Plains by 1,100 m. It consists largely of Mesoproterozoic granite, but is capped by silicified aeolian sandstones of the 180 million years old Etjo Formation. These are in the form of a 30 m caprock which gives the Gamsberg its famous tabular summit. The nature of the underlying granite slopes of the Gamsberg is described by Moon and Selby (1983).

The age of the escarpment is a matter of debate (Partridge and Maud 1987) and its form and persistence are variable. Its development, in common with other passive tectonic plate margins (such as eastern South America and Western India), is probably closely related to continental fragmentation and rifting, but the timing of escarpment formation, retreat and uplift is more contentious. Kempf (2010) identifies two main theories in the literature. The first is that the Great Escarpment is the erosional remant of the rift shoulder that developed and was uplifted as a marginal bulge through the break up of Gondwana, and which since then has been worn back by as much as 100 km to its present position by erosional processes. This might have been enhanced by the continent rising up like a cork as material was eroded from the land and deposited in the ocean. This process is called 'erosional isostasy'. Attempts to model the evolution of the Great Escarpment in terms of rifting, denudation and isostasy are provided by Gilchrist et al. (1994) and Dauteuil et al. (2013).

The second hypothesis is that the escarpment came into being a long time after the rifting and is thus the outcome of more recent tectonic events. Testing these hypotheses relies upon having well-dated histories of uplift and erosion, as well as good theoretical models of isostatic behaviour. There remains considerable disagreement.

That considerable denudation has occurred since the Cretaceous is indubitable, but whether there were periods of major denudation or more consistent trends is more debated. The elevations of the Damaraland complexes (such as Brandberg and Erongo) above the surrounding plains suggests that well in excess of 1.5 km of denudation has occurred around these intrusions. Large amounts of Etendeka lavas have also been removed (Gilchrist et al. 1994).

The fact that Tertiary deposits such as the Tsondab Sandstone overlie the Namib plains suggests that a large amount of denudation occurred relatively soon after the early Cretaceous tectonic events. It is also conceivable, though not proven, that rates of denudation were reduced by progressive Cainozoic aridification (Gilchrist et al. 1994, p. 12220).

On Southern Africa's south-western margin as a whole there is some evidence from apatite fission track analysis and the offshore sedimentary record that the early Cretaceous was a time of rapid denudation and offshore sedimentation (Brown et al. 1990; Rust and Summerfield 1990; Gallagher and Brown 1999). This could imply that following continental fragmentation there was indeed a phase of early and rapid uplift and scarp erosion, which led to the stripping of large expanses of Karoo strata (Gilchrist and Summerfield 1994; Van der Wateren and Dunai 2001) and thus supports the first hypothesis. Comparative fission track and cosmogenic data assembled by Cockburn et al. (2000) also suggest low rates of denudation affecting the central Namib over the last 36 million years. Sediment budgets constructed by calculating amounts of sediment deposited on the ocean floor offshore from Namibia and South Africa confirm this, demonstrating that rates were high in the Lower Cretaceous but have been considerably lower during the Tertiary and Quaternary (Guillocheau et al. 2010). However, sediments from offshore of the Orange mouth suggest that in addition to high rates of denudation during the post-rift Lower Cretaceous, there was another phase of rapid denudation in the Upper Cretaceous, 50 million years after the rifting event, perhaps as a response to a significant rejuvenation of relief at that time (Rouby et al. 2009) thus supporting the second hypothesis.

Burke (1996, pp. 364–369) has also cast some doubt on whether the Great Escarpment is as old as some studies imply. He queried the idea that because the Great Escarpment is parallel to the rifted continental margin it necessarily formed at the same time. Moreover, under erosive rainfall regimes he suggested it could still be evolving quickly. Burke suggested that much of the escarpment's development has taken place over the last 30 Ma, thus contributing evidence in support of the second hypothesis. On the other hand, cosmogenic nuclide studies by Bierman and Caffee (2001) indicated that there has been significant landscape stability over at least the past million years. Similarly, Van der Wateren and Dunai (2001), also using cosmogenic nuclides, found that long-term rates of denudation have been very low (c 5 m per million years), especially over the last 5 million years thus supporting the first hypothesis. However, Codilean et al. (2012) have pointed out that the rates obtained by this method depend to a substantial degree on the grain size characteristics of the material that is sampled and that results based on pebble-sized clasts may underestimate palaeo-denudation rates. So, at present neither hypothesis can be ruled out until better dating evidence is available from a wider range of locations.

Another important question that needs to be answered is why the Great Escarpment is relatively ill-developed over much of the Central Namib (Birkenhauer 1991) and why there is what Kempf (2010) called 'The Escarpment Gap'. One possible reason is that it is traversed by five rivers, including the Swakop and the Ugab, that have catchments that are longer and larger than those to the north and south. They derive power and discharge from areas that have relatively high rainfall and so may have concentrated erosion on this zone (Gevers 1936).

Alternatively, Hüser (1989) postulated that the gap is essentially lithological in origin, and is caused on the one hand by the belt of Cretaceous Damara granites and on the other by the fact that the well-developed escarpment to the north and south is the result of the presence of resistant Permo-Triassic Karoo sediments and the remains of Cretaceous Etendeka lavas. The granites, he argues, are not part of the escarpment but ought to be considered "foreign objects" which do not develop steps and terraces and are as such not able to form a classic escarpment. Spönemann and Brunotte (1989), on the other hand, favoured tectonism as being responsible for the break of the escarpment in the Central Namib Desert. They sought the cause in continental and regional scale deformations. The argument rumbles on, and it is still not clear which explanation is correct.

1.4 The Rivers

Namibia is a predominantly dry nation, and its rivers reflect this fact. On its borders there are three perennial rivers that reach the sea, the Kunene, the Zambezi and the Orange. Two other perennial rivers, the Okavango and the Kwanda flow into the inland Okavango Delta and the Linyanti Swamps of northern Botswana. All these rivers gain their flow from relatively humid and mountainous areas in Angola, Zambia and South Africa. The rivers that rise in Namibia itself are all ephemeral and seasonal. The great majority are dry for most of the year. Some, such as the Kuiseb, may not flow for several years and only occasionally reach the sea, while others, such as the Tsondab (Stone et al. 2010) and Tsauchab never manage to reach it under current conditions, terminating in the Tsondab and Sossus vleis respectively.

Details of the flows and catchment areas of the Namibian river catchments are provided by Strohbach (2008). The following are the largest catchments (areas in km^2):

Omatako Omuramba	61057
Etosha	57030
Fish	54326
Orange	44068
Nossob	37904
Ugab	29355
Auob	24540
Swakop	21010
Cuvelai	20730

However, there is some variability in estimates of catchment extent and, for example, Jacobson et al. (1995) estimate the area of the Swakop to be 30,100 km^2 and that of the Ugab to be 28,400 km^2. In the following paragraphs we discuss the nature of the perennial and ephemeral river systems of Namibia, starting with the ephemeral river systems that drain into the Atlantic, followed by the Orange and its tributaries, and ending with the internally-draining (endoreic) river systems of eastern Namibia.

The ephemeral rivers of the Namib, between the Orange in the south and the Kunene in the north (Jacobson et al. 1995; Jacobson and Jacobson 2013), rise in the interior mountains of Namibia and when they flow it is generally because of summer storms. In the south the Tsondab and the Tsauchab empty into pans within the Namib Sand Sea, and sometimes cause them to flood. In the past they flowed further west (Stone et al. 2010) and the Tsondab may have reached the Atlantic early in the Pleistocene (Seely and Sandelowsky 1974). The Tsauchab is notable for the way it has incised down into fan gravels and Tsondab Sandstone to produce the Sesriem Canyon (Grünert 2000, p. 133).

The Kuiseb, which has a catchment area of 15,500 km^2 and a length of c 420–440 km, rises in the Khomas Hochland of central Namibia and becomes incised into a canyon tract (Huntley 1985), which is up to 250 m deep and 1,000 m wide at its deepest part some 100 km from the coast. Incision occurred in the late Neogene (Van der Wateren and Dunai 2001). Dating and analysis of flood deposits indicate that this tract has received some 35 major floods over the last 1,300 years, with one attaining a discharge of 1,350 m^3/s (Grodek et al. 2013). The river is lined by various terrace fragments, including the Pleistocene Oswater Conglomerate and the Gobabeb Gravel Formation (Ward 1987). Lower down its course, the Kuiseb becomes a braided, sandy alluvial channel, forms the northern boundary of the Namib Sand Sea (Fig. 1.7), except at its most seaward point, and disappears into a delta behind Walvis Bay. It only reached the sea three times in the Twentieth Century, in 1933, 1962–1963, and 1985. However, it welcomed the new millennium by flowing to the coast in 2000 and it flooded the salt works in Walvis Bay in 2011. These floods stop the northward movement of dunes and help to account for the largely sand-free nature of much of the Central Namib plains. In flood, it also carries prodigious amounts of woody debris (Jacobson et al. 1999) (Fig. 1.8). When it does flow, its discharge decreases rapidly downstream because of transmission losses into its bed. At Gobabeb (some 50 km from the coast) the volume of an event that statistically occurs one year in ten (known as the 10 year event) is 10 million m^3, whereas at Swartbank (around 30 km from the coast) it is 3 million m^3, and at Rooibank (c 20 km from the coast) a mere 0.15 million m^3. The same applies to peak discharges which for a 10 year event are 90 m^3/s at Gobabeb, 25 m^3/s at Swartbank, and 0.9 m^3/s at Rooibank (Heidbüchel 2007). Between Natab and the sea there is a series of buried palaeochannels of the Kuiseb, the southernmost of which enters the Atlantic just north of Sandwich Harbour (Klaus et al. 2008).

To the north of the Kuiseb there is a small river which fails to reach the sea. This is the Tumas (Wilkinson 1990). In turn, to the north of that is the Swakop River, with a length of 460 km, which rises in the Khomas Hochland and enters the Atlantic at Swakopmund. Like the Kuiseb it is prone to flood sporadically and in 1931 it demolished the railway bridge linking that town with Walvis Bay (Figs. 1.9 and 1.10). As a result of another flood event in 1934 it transported sediment which built the coastline out by over 1 km (Massmann 1983). More recently, the Swakop reached the sea in February 2009 and in March 2011. It carries a substantial sediment load, thus justifying its local name, which derives from the local Nama words *Tsoa* (anus) and *Xou* (excrement). The major tributary of the Swakop is the Khan. This has its origin near the settlement of Otjisemba north-west of Okahandja. From there the river course passes westwards to the town of Usakos, and has its confluence with the Swakop 40 km east of Swakopmund.

Fig. 1.7 The Kuiseb at Gobabeb, 2010. It passes through a snaking green line of riparian forest

Fig. 1.8 Woody flood debris in the Kuiseb bed, Gobabeb, 2011

The Swakop canyon contains some interesting minor landforms. For example, just 8 km upstream from the Khan confluence, there is a large spring tufa, with a high halite content which is notable for enveloping and preserving ostrich (*Struthio camelus*) feathers. The present course of the Swakop runs further north than it did in the past for there is sedimentological evidence that it used to flow into the present Tumas, located between Walvis Bay and Swakopmund (Van der Wateren and Dunai 2001).

The Omaruru River, the bed of which is an important aquifer, reaches the Atlantic just to the north of Henties Bay (Stengel 1966). It has a catchment area of c 13,100 km^2 and

Fig. 1.9 The Swakop bridge during the 1931 flood (from Digital Namibian Archive)

Fig. 1.10 The remains of the same bridge at the mouth of the Swakop which was demolished by a flood in 1931

a length of c 330 km. Like the larger Ugab it has a history of flooding. The Ugab is flanked by three conglomerate terraces composing the Bertram Conglomerate Formation at 160, 100 and 30 m above the modern river (Mabbutt 1951; Grünert 2000, p. 69) of which the famous Vingerklip (Fingerklip), located c 80 km southwest of Outjo and 45 km west of Khorixas, is an erosional remnant (Fig. 1.11).

Further north, other ephemeral rivers enter the Skeleton Coast (see Chap. 8) and one of these, the Hoarusib, has a suite of major silt terraces that formed as a result of flood flow being ponded up behind a dune cordon. It reaches the sea most years. The Uniab, Hoanib and Hunkab are other rivers that are periodically ponded up by the Skeleton Coast erg (another name for a sand sea). Many of these rivers have comparatively lush riparian vegetation along their channels, a stark contrast to the adjacent sand and rock desert.

The Kunene rises in the Bié highlands of Angola, where it is called the Cunene, and flows into the Atlantic on the border between Angola and Namibia (Nicoll 2010). For much of its course it flows southwards, as if towards the Etosha Pan, an ancient structural basin (Buch and Trippner 1997), but then it turns sharply westwards and enters a tract

Fig. 1.11 Google Earth image of a terrace remnant on the Ugab River. Scale bar 0.25 km (© 2013 Digital Globe)

with steep falls and rapids (e.g. the Caxambue rapids and the Epupa and Ruacana Falls). The Ruacana Falls are c 120 m high, while at Epupa the river forms a series of riffles and cascades that drop a total of around 60 m over c 1.5 km. Good photos of the river at Ruacana prior to dam construction are provided by Kanthack (1921). Between Ruacana and the Atlantic the altitude of its bed drops by more than 1,100 m over a distance of 340 km. These characteristics seem to indicate that this is a case of a river capture by a stream eroding backwards from the coast and capturing the interior drainage (Wellington 1955, p. 65). Wellington also suggests that the conditions for a similar process of capture are present in the headwaters of the Rio Coroca to the north of the lower Kunene. This river, having eroded headward through the Sierra de Chela of the Great Escarpment, is threatening to behead the upper Caculuvar River, a tributary of the upper Kunene.

The timing of the Kunene capture is not well constrained (Moore and Blenkinsop 2002) but lowering of the base-level associated with the opening of the Atlantic may have initiated a period of rapid erosion, which may have exploited a Permo-Carboniferous glacial valley from which Karoo sediments were stripped. Miller (2008, Chap. 24) suggests that the capture of the Kunene took place towards the end of a widespread phase of drainage incision during the Miocene-Pliocene. It has also been argued that the capture, which perhaps caused the shrinkage of a postulated large Lake Etosha, occurred only c 35,000 years ago (Buch 1997). At the far western end of its course, as it passes through the coastal sand sea, there are signs that the Kunene formerly entered the sea considerably to the south of its present mouth (Sander 2002), and that it may have been forced northwards by dune encroachment. At Serra Cafema the Kunene has a 5 m high terrace with a sprinkling of unremarkable Middle Stone Age flakes on its surface (Nicoll 2010).

The Okavango is a perennial, endoreic river with ephemeral tributaries (Seely et al. 2003; Strohbach 2013). It rises in the south-western Angolan highlands, near and just east of the source of the Kunene and Cuvelai rivers. It flows for more than 600 km from the upper catchment in a southerly direction until it reaches border between Angola and Namibia. From that point, the river forms the border between Angola and Namibia over a distance of some 400 km. Its channel, which lies c 40–60 m lower than the surrounding sand plateau, is characterised by scroll bars, abandoned channels, oxbow lakes etc. (Fig. 1.12). The river is notable for the fact that it carries a very low silt and clay load, with the bulk of its sediment appearing to be re-worked fine aeolian sand. This is transported as bed load and creates sub-aqueous dune bedforms. Just before it enters its panhandle in Botswana, the river crosses the Popa Falls, which in the dry season have a visible height of c 3.5 m. The month of maximum flow is April. In the far east of the Caprivi, the Zambezi forms the border with Zambia. The Caprivi is often subject to severe flooding (Skakun et al. 2013).

The Cuvelai River is also an endoreic river, rising in the southern foothills of the Sierra Encoco in southwestern Angola (Mendelsohn and Weber 2011). It drains southwards

Fig. 1.12 Google Earth image of the Okavango Floodplain. Note the oxbow lakes, abandoned channels, point and scroll bars, etc. Scale bar 0.5 km (© 2012 Google, US Dept. of State Geographer, Digital Globe)

towards the Etosha Pan and is perennial for about 100 km before it ramifies into a delta of ephemeral watercourses (Lindeque and Archibald 1991) which cross a broad plain of low relief; this delta converges again to terminate in the ephemeral Etosha pan. The watercourses (Fig. 1.13), called oshanas, appear to be a complex network of flood channels, most of which are oriented from northwest to southeast. They are the lifeblood of an area where just less than half of the population in Namibia live. However, severe floods can cause many deaths as in March 2009 and March 2011.

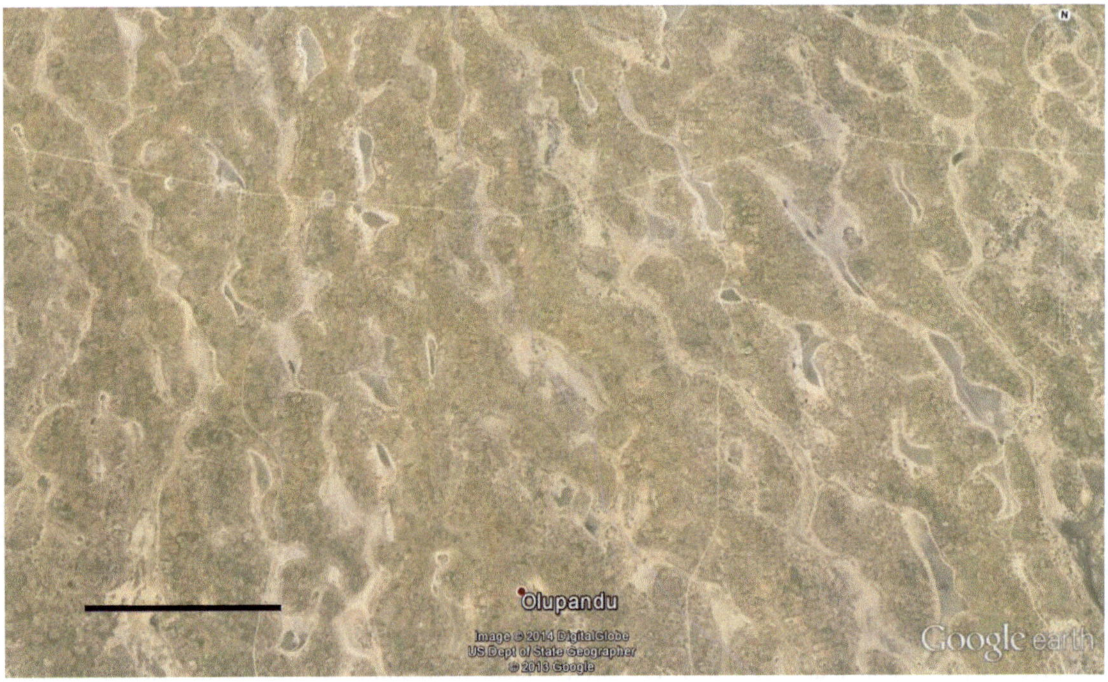

Fig. 1.13 Google Earth image of Oshanas. Scale bar 5 km (© 2014 Digital Globe)

The perennial Orange River, which occurs on the southern border of Namibia, originates in the Maloti Highlands of north-eastern Lesotho and after flowing through regions of steadily increasing aridity reaches the South Atlantic at Alexander Bay (Bluck et al. 2007). It is the largest catchment in southern Africa, and about one quarter of its area is within Namibia. Although its flow is now heavily regulated by dams, the river has been subject to major floods. A peak in flood activity may have occurred around 500 years ago in a phase contemporary with the Little Ice Age (Heine and Völkel 2011). The sediment load of the Orange, now greatly reduced compared to natural levels because of entrapment in reservoirs, has included diamond-rich gravels (Corbett and Burrell 2001; Spaggiari et al. 2006), and it has also been a major source of sand for the dunes of the Namib Sand Sea (Garzanti et al. 2012). Its terraces include the Arries Drift Gravel Formation, (which is 19–17 Ma old), also known as the Proto-Orange terrace, and a lower Meso-Orange terrace, which is thought to be of Plio-Pleistocene age (Jacob et al. 1999; Jacob 2005). A map of these terraces appears in Miller (2008, Fig. 25.16). The mouth of the Orange River is an example of a delta dominated by wave action and longshore drift. Rather than accumulating at the river mouth as a classic and visible delta, the sediments have been carried up the coast and onshore by the strong swells and onshore winds.

Box 1: The Fish River and its world class canyon

The Fish River rises in Namaqualand and flows south across the Great Namaqualand plateau, where it cuts a spectacular gorge or canyon before emptying into the Orange River. It is about 600 km long and is intermittent. The river often flows in the summer months (especially January to March) and severe flash floods can pose problems for the tourist resort at Ai-Ais. In 1973/74 and 1975/76, two very wet seasons, flows at the Hardap Dam exceeded 6,000 m^3/s. Its Kam, Schlip and Kalf tributaries originate in the central highland area south of Rehoboth whilst the Narub and Usib Rivers flow from the eastern foothills of the Naukluft Mountains. The Hutup, Lewer and Kanibes Rivers drain from the northern and eastern parts of the Schwarzrand Mountains. The Löwen and Gaub Rivers originate in the Groot Karas Mountains and the Konkiep in the western Schwarzrand. It is possible that prior to its present course, the Fish flowed southeastwards from its southerly bend at Ganikobis past Tses and in the direction of the line of pans extending along the sandy lowland towards Aroab (Wellington 1967, p. 26).

The Fish River canyon is one of the greatest spectacles in Namibia (Fig. 1.14). Its scale is stupendous. It is often said to be the second greatest canyon in the world after the very much larger Grand Canyon of the Colorado in the USA. Whether this is true is a matter of debate for the Blue Nile gorge in Ethiopia is very much deeper. Other enormous canyons include the Tsangpo in Tibet and the Copper Canyon in Mexico. For the first 450 km of its course the Fish has a limited gradient and flows within a broad valley. However, after its confluence with the Löwen River, it

Fig. 1.14 The Fish River Canyon

begins to become incised and eventually enters a canyon tract (Simpson and Davies 1957). This is located about 80 km west of Grünau, and starts about 30 km upstream of the Ai-Ais hot springs, and extends for about 50 km. The gorge is between 160 and 550 m deep, and 5–8 km wide, and is incised into flat-lying Nama sediments and into the underlying Namaqua Complex gneisses, which themselves are some 1,800 million years old. The river, in incising, has created some enormous entrenched meanders. The river must initially have flowed over a flat land surface where it could develop its bends freely. Then, continental uplift associated with the breakup of Gondwanaland in the Lower Cretaceous occurred and this was a major factor that caused the incision to occur. The canyon is also associated with some major fault and graben structures and is thus in part a rift valley (Mvondo et al. 2011; Kounov et al. 2013). It has been of particular interest recently as an analogue of some of the great valley networks on Mars (Petau et al. 2011).

The endoreic Kalahari catchments, such as the Omatako Omurambo, are prone to lose much of their discharge in the Kalahari sands. Some of them extend into Botswana and across to the margins of the Okavango Delta and Lake Mkgadikgadi. These are the dry valleys or *mekgacha* systems and are thought to be relicts of formerly more humid conditions (Thomas and Shaw 1991, p. 136). Examples include the Okwa and the Groot Laagte. The Auob and the Nossob, partially incised into calcreted valley sides, negotiate the linear dunes and eventually flow southeastwards into the Molopo. The Nossob has its origin in two main tributaries, the Swart-Nossob and Wit-Nossob, meaning black and white respectively. Both tributaries have their origins in the eastern slopes of the Otjihavera mountain range, east of Windhoek. Their sources are at 1,800 m and over 2,000 m above sea level respectively. The two river beds have their confluence some 80 km south of Gobabis, which is situated on the bank of the Swart Nossob.

One of the most intriguing features of the Namib rivers is that many of them, in contrast to most 'normal' rivers in temperate climates, display convex long profiles over all or much of their courses. This is the case for the Kunene, Kuiseb, Omaruru, Swakop, Tumas, and Ugab (Dauteuil et al. 2013), as well as some of the southern Angolan rivers, and the lower course of the Orange. Whether this is due to the nature of the uplift on this passive tectonic continental margin, or to the fact that in this arid environment river discharges diminish downstream, or to a combination of both, is a matter of debate. The traditional explanation is that in dryland rivers there is a diminution in flow downstream because of transmission losses, and so the ratio of sediment to flow often increases downstream, leading to aggradation and the development of a convex profile. It is also possible that in these rivers there is a less clear diminution in grain size of sediment downstream in comparison with humid climate rivers. Certainly, studies of drainage basins in the high plains of the USA (Zaprowski et al. 2005), indicate that in tectonically stable settings, areas with higher intensity rainfall and greater mean annual precipitation, have increasingly concave long profiles. However, a strong case has been made that uplift rate histories explain many of the main characteristics of the long profiles of those rivers that drain the tectonic swells of Africa (Roberts and White 2010).

1.5 The Coastline

The coastline of Namibia stretches 1,570 km from the mouth of the perennial Kunene river in the north to that of the perennial Orange river in the south (Bird et al. 2010). In between, all the rivers that flow into the Atlantic are ephemeral. The coastal environments of Namibia are shaped by the dynamic interplay between sediments coming from on land (transported by rivers and the wind) and the waves, tides and currents which characterise the eastern margins of the Atlantic Ocean. Namibia is located within a swell wave environment, and experiences near constant, and often large, waves coming from the SW which have travelled huge distances across the Atlantic. Namibia's tidal regime is categorised as microtidal, with tidal ranges at Walvis Bay in the order of 1 m. Mean annual tidal range is roughly equivalent to the mean significant wave heights, meaning that both wave and tidal processes are crucial to shaping the coastal environment. The Namibian coast also experiences the effects of the Benguela Current, a strong surface current which comes from the Southern Atlantic and brings cold waters up the coast to around the mouth of the Kunene River. It is driven by Southern Atlantic anticyclonic winds which are strongest in winter. Counter currents flow south closer to the shore, and at depth below the Benguela Current. Along the Namibia Coast there are also coastal areas of surface upwelling (the largest of which is at Lüderitz) where cold water ascends from depth. These upwelling areas support important fisheries. The Namibian coast is fronted by continental shelf—which extends some 100 km off most of the coastline, narrowing to 35 km wide north of the Walvis Ridge. Details of the coastal sediments are provided by Rogers and Rau (2006), and the distribution of the main coastal types is discussed by Harris et al. (2012), who note that the predominant coastal form is the sandy beach, making up about 68.5 % of the total. Other important coastal landscapes are rocky coasts, deltas and complex spit and lagoon systems.

The Skeleton Coast southwards from the Kunene mouth (which is obstructed by a bar), is backed by a series of major salt pans or coastal sabkhas, which have been mapped by Schneider and Genis (1992). Their origin is uncertain and deserves further attention. At Cape Fria low hills of Cretaceous basalt end in cliffs (Noli and Avery 1987). Between there and Möwe the shore is sandy with fringing rock reefs at False Cape Fria. Rocky Point is another prominent basalt headland, whereas Möwe is a low foreland of Damara metasediments. From Möwe to Palgrave Point the low lying coast is backed by scattered salt pans and small shifting dunes. Damara metasediments and Cretaceous basalts also outcrop. With varying success a series of rivers cut through the dune cordon and the beach, but some are ponded up (see Chap. 8). From Palgrave Point to Cape Cross there are alternations of rocky cliffs, again in Damaran metasediments and Cretaceous basalts, and there are many salt pans. Cape Cross itself projects out about 5 km into the Atlantic and is the site of a large, smelly and noisy seal colony. The coast from Cape Cross southwards is generally rocky until c 50 km north of Walvis Bay. There are marine terraces at c 2–17 m above storm tide level (Wieneke and Rust 1973) and some extending upwards to a maximum of 30 m (Davies 1973). At Swakopmund the mouth of the Swakop River has from time to time built outwards as a result of flood flows, which have, as we have already seen, also demolished the old bridge between Swakopmund and Walvis Bay. The next important river is the Kuiseb, which forms a delta just to the south of Walvis Bay (Huntley 1985). The town is an important lagoonal harbour, protected by the Walvis Peninsula and Pelican Point. Tides along this coastline are regular and semi-diurnal. There is a mean spring tide range of 1.42 m (0.27–1.69 m) and a mean neap tidal range of 0.62 m (0.67–1.29 m). These semi-diurnal tides flush Walvis Bay twice daily.

Buffetted as it is by large waves, strong winds, and rapid long-shore drift of sediment, the southern coast of Namibia is a highly dynamic environment with ever-changing spits and lagoons (Watson and Lemon 1985). One of these spit and lagoon systems is that at Walvis Bay. Strong longshore sediment transport from the south drives the Walvis Peninsula spit northwards at a rapid rate (Elfrink et al. 2003). The dominant wave direction is between 225° and 270° (Hughes et al. 1992) and wave data are given in detail in www.gecko.na/documents/ffd_o5a.pdf (accessed January 19th 2013). Analysis of old maps and photographs showed that its tip, Pelican Point, grew by an average of 17.4 m per year between 1885 and 1980. Between 1980 and 1996 Pelican Point prograded over a distance of 340 m, an average of 22.6 m per year (Schoonees et al. 1998). In all, the spit extended northwards by some 760 m between 1973 and 2010. From time to time, as in 1900 (Waldron 1900), 1959 and 2000, ephemeral mud islands have developed offshore and have been associated with gas eruptions (Emeis et al. 2004).

Box 2: The shifting sands of Sandwich Harbour

South of Walvis Bay is Sandwich Harbour (Fig. 1.15) (often called Sandvis and before that Porto D'Ilheo), which lies astride the Tropic of Capricorn. This convex sandplain is about 15 km long and protects a 9 km long lagoon, at the south end of which is a damp salt pan (Wilkinson et al. 1989). At the north end there is a highly unstable spit. In the nineteenth century Sandwich was a much used anchorage and port, visited by British and American whalers. Early maps of the

Fig. 1.15 Sandwich Harbour, August 2010

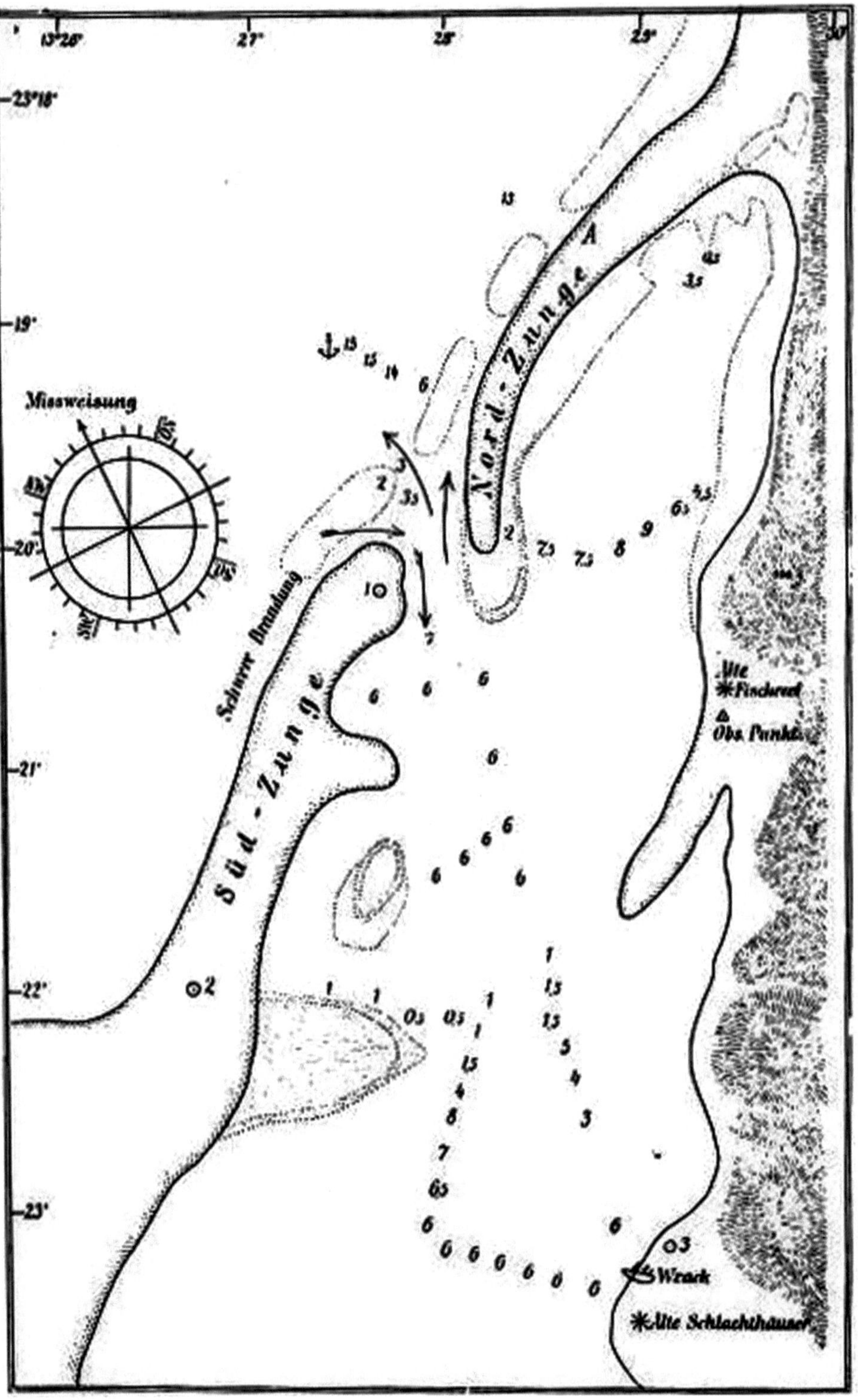

Fig. 1.16 Sandwich Harbour in 1905 (from Schultze 1907)

harbour for 1880, 1889 and 1905 are presented in Schultze (1907). The first two indicate that the entrance to the harbour may have been 9 m deep, but the last shows that it had been displaced southwards and had shallowed to only c 3 m (Fig. 1.16). Bar growth and silting since the late nineteenth century mean that it is no longer usable (Kensley and Penrith 1977). Note the rapid changes that took place in just a few decades as recorded by the satellite images in Fig. 1.17. Sandwich Harbour may mark a former

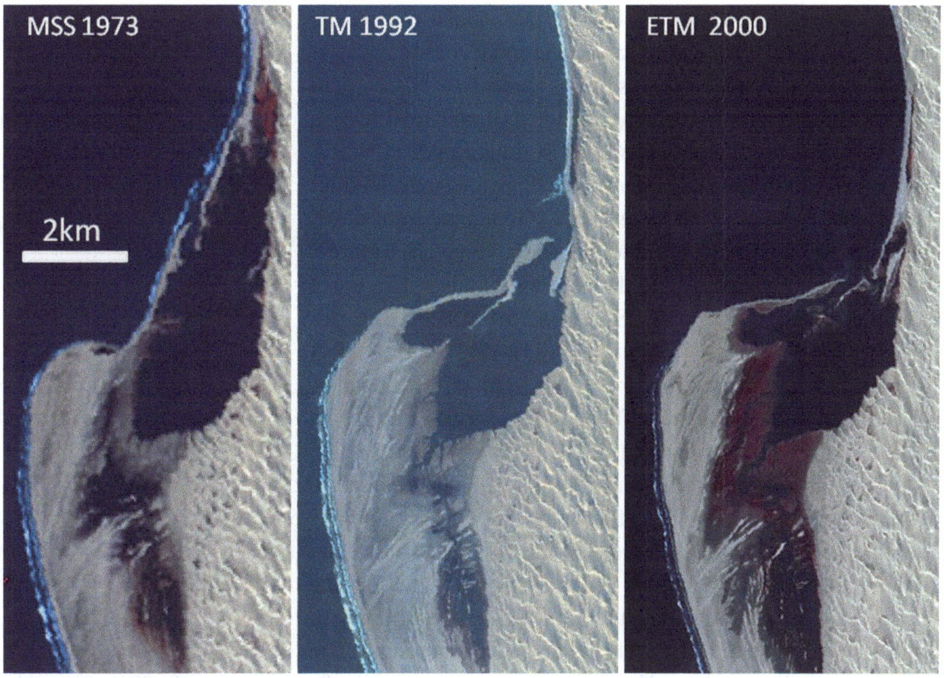

Fig. 1.17 A sequence of Landsat images from 1973 to 2000, showing changes in Sandwich Harbour (courtesy of Dr Frank Eckardt)

mouth of the Kuiseb, and the other headlands of this type may also be ancient river mouths produced by streams that formerly flowed across the Namib Sand Sea, but which now terminate in inland sumps like Sossus Vlei and Tsondab Vlei.

The third and fourth spit and lagoon systems are Conception and Meob Bays. At Conception Bay (Fig. 1.18) the degree of change is shown by the rusting remains of a German steamship, the *Eduard Bohlen*, which ran aground in 1909 (Harris et al. 2012). By 1973 the wreck was about 400 m inland from the shore. Here the spit has closed the

Fig. 1.18 Google Earth image of Conception Bay. Scale bar 2 km. Note the small enclosed pan at the northern end and the many old ridges at the southern end (© 2012 Digital Globe, TerraMetrics, Google)

lagoon completely to form a coastal pan, and it is possible that this will in due course be the ultimate destiny of the Sandwich and Walvis spits as well. Meob Bay has a small lagoonal pan at its northern end, but its main feature is a suite of ancient recurved spits at its southern end.

From Walvis Bay to Lüderitz the coastline is backed by the Namib Sand Sea and large dunes often come right down to the shoreline. South of Lüderitz, where there are bays cut into basement gneisses and schists of pre-Damara age, is the so-called Diamond Coast. South of Elizabeth Bay are beautifully developed barchan dunes, while at Bogenfels there are cliffs and a classic arch, c 55 m high, developed in Gariep Complex dolomites. There is also a Holocene raised beach (Compton 2006). North of Chameis Bay the coastline is roughly linear and has many rocky headlands and small, north-facing sandy bays, which are known as 'J-bays' because of their shape. The rocks that form these J-bays and other south-facing re-entrant bays belong to the Gariep Group. From Chameis Bay to the Orange, the straightish coastline has been intensively mined for diamonds. The coastline of the southern Namib also has many small offshore islands. Finally, at the border with South Africa is the mouth of the Orange River, a source of sediment for the coastline to its north (Bluck et al. 2007).

References

Beaudet G Michel P (1978) Recherches géomorphologiques en Namibie central. Recherches Géographiques à Strasbourg, Special Issue 139

Bertram S (2003) Late Quaternary sand ramps in south-western Namibia. Nature, origin and palaeoclimatological significance. Doctoral Dissertation, University of Würzburg

Besler H (1972) Klimäverhaltnisse und klimageomorphologische Zonierung der zentralen Namib (Südwestafrika). Stuttgarter Geographische Studien, Band 83

Besler H, Blümel W-D, Heine K, Hüser K, Leser H, Rust U (1994) Geomorphogenese und Paläoklima Namibias. Erde 125:139–165

Bierman PR, Caffee M (2001) Slow rates of rock surface erosion and sediment production across the Namib Desert and escarpment, southern Africa. Am J Sci 301:326–358

Bird E, Goudie A, Viles H (2010) Namibia. In: Bird E (ed) Encyclopedia of the world's coastal landforms. Springer, Dordrecht, pp 961–973

Birkenhauer J (1991) The Great Escarpment in southern Africa and its coastal forelands—a re-appraisal. Münchener Geogr Abh Reihe B. Band B11

Bluck BJ, Ward JD, Cartwright J, Swart R (2007) The Orange River, southern Africa: an extreme example of a wave-dominated sediment dispersal system in the South Atlantic Ocean. J Geol Soc 164:341–351

Brown RW, Rust DS, Summerfield MA, Gleadow AJW, de Wit MCJ (1990) An early Cretaceous phase of accelerated erosion on the south western margin of Africa: evidence from apatite fission track analysis and the offshore sedimentary record. Nucl Track Radiat Meas 17:339–350

Brunotte E, Sander H (2000) Loess accumulation and soil formation in Kaokoland (northern Namibia) as indicators of Quaternary climatic change. Glob Planet Change 26:67–75

Brunotte E, Maurer B, Fischer P, Lomax J, Sander H (2009) A sequence of fluvial and aeolian deposits (desert loess) covering the last 60 ka in the Opuwo basin (Kaokoland/Kunene Region, Namibia) based on luminescence dating. Quat Int 196:71–85

Bryant RG (2003) Monitoring hydrological controls on dust emissions: preliminary observations from Etosha Pan, Namibia. Geogr J 169:131–141

Buch MW (1997) Etosha Pan—the third largest lake in the world? Modoqua 20:49–64

Buch MW, Trippner C (1997) Overview of the geological and geomorphological evolution of the Etosha region, northern Namibia. Madoqua 20:65–74

Burke K (1996) The African plate. S Afr J Geol 99:341–409

Cloos H (1937) Südwestafrika. Geol Rundsch 28:163–187

Cockburn HAP, Brown RW, Summerfield MA, Seidl MA (2000) Quantifying passive margin denudation and landscape development using a combined fission-track thermochronology and cosmogenic isotope approach. Earth Plan Sci Letters 179:429–435

Codilean AT, Fenton CR, Fabel D, Bishop P, Xu S (2012) Discordance between cosmogenic nuclide concentrations in amalgamated sands and individual fluvial pebbles in an arid zone catchment. Quat Geochron. doi:10.1016/j.quageo.2012.04.007

Compton JS (2006) The mid-Holocene sea-level highstand at Bogenfels Pan on the southwest coast of Namibia. Quat Res 66:303–310

Corbett I (1993) The modern and ancient pattern of sandflow through the southern Namib deflation basin. Spec Publs Int Ass Sediment 16:45–60

Corbett I, Burrell B (2001) The earliest Pleistocene (?) Orange River fan-delta: an example of successful exploration delivery aided by applied Quaternary research in diamond placer sedimentology and palaeontology. Quat Int 82:63–73

Dauteuil O, Deschamps F, Bourgeois O, Mocquet A, Guillocheau F (2013) Post-breakup evolution and palaeotopography of the north Namibian Margin during the Meso-Cenozoic. Tectonophysics 589:103–115

Davies O (1973) Pleistocene shorelines in the western cape and South-West Africa. Ann Natal Mus 21:719–765

De Vries JJ, Selaolo ET, Beekman HE (2000) Groundwater recharge in the Kalahari, with reference to paleo-hydrologic conditions. J Hydrol 238:110–123

Dingle RV, Bremner JM, Giraudeau J, Buhmann D (1996) Modern and palaeo-oceanographic environments under Benguela upwelling cells off southern Namibia. Palaeogeogr Palaeoclim Palaeoecol 123:85–105

Eckardt F, Drake N (2011) Introducing the Namib Desert playas. In: Ozturk M et al (eds) Sabkha ecosystems. Springer, Heidelberg, pp 19–25

Eckardt F, Drake N, Goudie AS, White K, Viles H (2001) The role of playas in pedogenic gypsum crust formation in the Central Namib Desert: a theoretical model. Earth Surf Proc Land 26:1177–1193

Eckardt FD, Livingstone I, Seely M, von Holdt J (2013) The surface geology and geomorphology around Gobabeb, Namib Desert, Namibia. Geogr Ann 95A:271–284

Eitel B, Blumel WD, Hüser K, Mauz B (2001) Dust and loessic alluvial deposits in Northwestern Namibia (Damaraland, Kaokoveld): sedimentology and palaeoclimatic evidence based on luminescence data. Quat Int 76(77):57–65

Eitel B, Blümel WD, Hüser K (2004) Palaeoenvironmental transitions between 22 ka and 8 ka in monsoonally influenced Namibia. Lect Notes Earth Sci 102:167–194

Elfrink B, Prestedge G, Rocha CBM, Juhl J (2003) Shoreline evolution due to highly oblique incident waves at Walvis Bay, Namibia. In: Proceedings of intranational conference on coastal sediments
Emeis K-C, Brüchert V, Currie B, Endler R, Ferdelman T, Kiessling A, Leipe T, Noli-Peard K, Struck V, Vogt T (2004) Shallow gas in shelf sediments of the Namibian coastal upwelling system. Cont Shelf Res 24:627–642
Gallagher K, Brown R (1999) Denudation and uplift at passive margins: the record on the Atlantic Margin of southern Africa. Phil Trans R Soc Lond A 357(1753):835–859
Garzanti E, Andò S, Vezzoli G, Lustrino M, Boni M, Vermeesch P (2012) Petrology of the Namib Sand Sea long-distance transport and compositional variability in the wind-displaced Orange Delta. Earth-Sci Rev 112:173–189
Garzanti E, Vermeesch P, Padoan M, Resentini A, Vezzoli G, Andò S (2014) Provenance of passive-margin sand (Southern Africa). J Geol 122:17–42
Gevers TW (1936) The morphology of western Damaraland and the adjoining Namib Desert of South West Africa. S Af Geogr J 19:61–79
Gilchrist AR, Summerfield MA (1994) Tectonic models of passive margin evolution and their implications for theories of long-term landscape development. In: Kirkby MJ (ed) Process models and theoretical geomorphology. Wiley, Chichester, pp 55–84
Gilchrist AR, Kooi H, Beaumont C (1994) Post-Gondwana geomorphic evolution of south western Africa: implications for the controls of landscape development from observations and numerical experiments. J Geophys Res 99(B6):12111–12228
Goudie AS, Eckardt F (1999) The evolution of the morphological framework of the central Namib Desert, Namibia, since the early Cretaceous. Geogr Ann 81A:443–458
Goudie AS, Thomas DSG (1985) Pans in southern Africa with particular reference to South Africa and Zimbabwe. Z Geomorph 29:1–19
Grodek T et al (2013) The last millennium largest floods in the hyperarid Kuiseb River basin, Namib Desert. J Quat Sci 28:258–270
Grünert N (2000) Namibia. Fascination of Geology. Klaus Hess, Windhoek
Guillocheau F et al (2010) Quantification and causes of the terrigenous sediment budget at the scale of a continental margin: a new method applied to the Namibia-South Africa margin. Basin Res 24:3–30
Haddon IG, McCarthy TS (2005) The Mesozoic-Cenozoic interior sag basins of Central Africa: the Late-Cretaceous-Cenozoic Kalahari and Okavango basins. J Afr Earth Sci 43:316–333
Harris L, Holness S, Nel R, Lombard AT, Schoeman D (2012) Intertidal habitat composition and regional-scale shoreline morphology along the Benguela coast. J Coast Conserv 17:143–154
Heidbüchel I (2007) Recharge processes in ephemeral streams derived from a coupled stream flow routing/groundwater model—application to the Lower Kuiseb (Namibia). Thesis, Freiburg University
Heine K, Völkel J (2011) Extreme floods around AD 1700 in the northern Namib Desert, Namibia, and in the Orange River catchment, South Africa—were they forced by a decrease of solar irradiance during the Little Ice Age? Geogr Polonica 84:61–80
Hövermann J (1978) Formen und Formung in der Pränamib (Flächen-Namib). Z Geomorph Supp Bd 30:55–73
Hughes P, Brundrit GB, Searson S (1992) The vulnerability of Walvis Bay to rising sea levels. J Coastal Res 8:868–881
Huntley BJ (ed) (1985) The Kuiseb environment: the development of a monitoring baseline. South African National Scientific Programmes Report, 106
Hüser K (1989) Die Südwestafrikanische Randstufe. Grundsätzliche Probleme ihrer geomorphologischen Entwicklung. Z Geomorph Suppl 74:95–110
Jacob RJ (2005) The erosional and Cainozoic depositional history of the Lower Orange River, southwestern Africa. PhD thesis, University of Glasgow
Jacob RJ, Bluck BJ, Ward JD (1999) Tertiary-age diamondiferous fluid deposits of the lower Orange River valley, Southwest Africa. Econ Geol 94:749–758
Jacobson PJ, Jacobson KM (2013) Hydrologic controls of physical and ecological processes in Namib Desert ephemeral rivers: implications for conservation and management. J Arid Env 93:83–93
Jacobson PJ, Jacobson KM, Angermeier PL, Cherry DS (1999) Transport, retention, and ecological significance of woody debris within a large ephemeral river. J N Am Benthol Soc 18:429–444
Jacobson PJ, Jacobson KM, Seely MK (1995) Ephemeral rivers and their catchments. Desert Research Foundation of Namibia, Windhoek
Kanthack FE (1921) Notes on the Kunene River, southern Angola. Geogr J 57:321–336
Kempf J (2010) Morphotectonics and denudation in the landform history of the Namibian Great Escarpment landscapes. Z Geomorph 54:347–376
Kensley BF, Penrith MJ (1977) Biological survey of Sandvis 1, introduction and faunal list. Modoqua 10:181–190
Klaus J, Külls C, Dahan O (2008) Evaluating the recharge mechanism of the Lower Kuiseb Dune area using mixing cell modelling and residence time data. J Hydrol 358:304–316
Kounov A, Viola G, Dunkl I, Frimmel HE (2013) Southern African perspectives on the long term morpho-tectonic evolution of cratonic interiors. Tectonophysics 601:177–191
Lageat Y (1994) Le désert du Namib central. Ann Géogr 103:339–350
Lancaster N (1982) Dunes of the Skeleton Coast, Namibia (South West Africa): geomorphology and grain-size relationships. Earth Surf Proc Land 7:575–587
Lancaster N (1984) Characteristics and occurrence of wind erosion features in the Namib Desert. Earth Surf Proc Land 9:469–478
Lancaster N (1989) The Namib Sand Sea: dune forms, processes and sediments. Balkema, Rotterdam
Lindeque M, Archibald TJ (1991) Seasonal wetlands in Owambo and the Etosha National Park. Madoqua 17:129–133
Logan RF (1960) The Central Namib Desert, South West Africa. National Academy of Sciences/National Research Council Publication, 758, Washington DC
Mabbutt JA (1951) The evolution of the Middle UgabValley, Damaraland, South West Africa. Trans R Soc S Afr 33:333–363
Mabbutt JA (1952) A study of granite relief from South-West Africa. Geol Mag 89:87–96
Maier WD, Rasmussen B, Fletcher IR, Li C, Barnes SJ, Huhma H (2013) The Kunene Anorthosite Complex, Namibia, and its satellite intrusions: geochemistry, geochronology, and economic potential. Econ Geol 108:953–986
Massmann U (1983) Swakopmund. Society for Scientific Development and Museum, Swakopmund
Mendelsohn J, Jarvis A, Roberts C, Robertson T (2002) Atlas of Namibia. Sunbird Publishers, Cape Town
Mendelsohn J, Weber B (2011) The Cuvelai basin: its water and people in Angola and Namibia. www.raison.com.na/CUVELAI-Angola-Namibia atlas final_low.pdf (accessed 15 Apr 2013)
Miller RMG, Pickford M, Senut B (2010) The geology, palaeontology and evolution of the Etosha Pan, Namibia: implications for terminal Kalahari deposition. S Afr J Geol 113:307–334
Miller RMcG (2008) The geology of Namibia, vol 3. Geological Survey of Namibia, Windhoek
Moon BP, Selby MJ (1983) Rock mass strength and scarp forms in southern Africa. Geogr Ann 65A:135–145

Moore A, Blenkinsop T (2002) The role of mantle plumes in the development of continental-scale drainage patterns: The southern African example revisited. S Afr J Geol 105:353–360
Mvondo F, Dauteuil O, Guillocheau F (2011) The Fish River canyon (southern Namibia): a record of Cenozoic mantle dynamics? C R Geoscience 343:478–485
Nash DJ, Shaw PA, Thomas DSG (1994) Duricrust development and valley evolution: process landform links in the Kalahari. Earth Surf Proc Land 19:299–317
Nicoll K (2010) Geomorphic development and middle stone age archaeology of the Lower Cunene River, Namibia-Angola border. Quat Sci Rev 29:1419–1431
Noli D, Avery G (1987) Stone circles in the Cape Fria area, northern Namibia. S Afr Archaeol Bull 42:59–63
Pallett J (ed) (1995) The Sperrgebiet. NAMDEB, Windhoek
Partridge T, Maud R (1987) Geomorphic evolution of southern Africa since the Mesozoic. S Afr J Geol 90: 179–208
Partridge TC, Maud RR (eds) (2000) The Cenozoic of southern Africa. Oxford University Press, New York
Passarge S (1904) Die Kalahari. Reimer, Berlin
Petau A, Tirsch D, Jaumann R (2011) Martian valley networks and analogue features on Earth. EurPlan Sci Cong Abstracts, 6, EPSC-DPS2011-830
Richthofen F von (1882) On the mode of origin of the loess. Geol Mag 9, series 2:293–305
Roberts GG, White N (2010) Estimating uplift histories from river profiles using African examples. J Geophys Res 115, B02406: doi:10.1029/2009JB006692
Rogers J, Rau AJ (2006) Surficial sediments of the wave-dominated Orange River Delta and the adjacent continental margin off south-western Africa. Afr J Marine Sci 28:511–524
Rouby D, Bonnet S, Guillocheau F, Gallagher K, Robin C, Biancotto F, Dauteuil O, Braun J (2009) Sediment supply to the Orange sedimentary system over the last 150 My: an evaluation from sedimentation/denudation balance. Marine Petrol Geol 26:782–794
Rust DJ, Summerfield MA (1990) Isopach and borehole data as indicators of rifted margin evolution in south western Africa. Marine Petrol Geol 7:277–287
Sander H (2002) Relief- und Regolithgenese im nordöstlichen Kaokoland/Namibia. Thesis, Universität zu Köln
Schneider GIC, Genis G (1992) Salt. Namibian Ministry of Mines and Energy, Geological Survey of Namibia, Windhoek, Open File Report MRS 10
Schoonees JS, Lenhoff L, Raw AJ (1998) Preventing natural breaching of the major sand spit protecting the port of Walvis Bay. Proc Coast Eng Conf 2:1475–1488
Schultze L (1907) Aus Namaland und Kalahari. Gustav Fischer, Jena
Seely M, Henderson J, Heyns P, Jacobson P, Nakale T, Nantanga K, Schachtschneider K (2003) Ephemeral and endoreic river systems: relevance and management challenges. In: Turton A, Ashton P, Cloete E (eds) Transboundary rivers, sovereignty and development: Hydropolitical drivers in the Okavango River basin. AWIRU and Green Cross International, Pretoria, pp 187–212
Seely MK, Sandelowsky BH (1974) Dating the regression of a river's end point. S Afr Arch Soc Goodwin Ser 2:61–64
Shaw AI, Goudie AS (2002) Geomorphological evidence for the extension of the Mega-Kalahari into south-central Angola. S Afr Geogr J 84:182–194
Shaw AI, Goudie AS (2004) Chemical sediments of the palaeo and active Okavango alluvial fan, Botswana: evidence for variations in the processes of formation. Z Geomorph Suppl 136:89–112
Shaw PA, Thomas DSG (1988) Lake Caprivi: a late Quaternary link between the Zambezi and middle Kalahari drainage systems. Z Geomorph 32:329–337
Simpson EJW, Davies DH (1957) Observations on the Fish River Canyon in South West Africa. Trans R Soc S Afr 35:97–107
Skakun S, Kussul N, Shelestov A, Kussul O (2013) Flood hazard and flood risk assessment using a time series of satellite images: a casestudy in Namibia. Risk Anal. doi:10.1111/risa.12156
Spaggiari RI, Bluck BJ, Ward JD (2006) Characteristics of diamondiferous Plio-Pleistocene littoral deposits within the palaeo-Orange River mouth, Namibia. Ore Geol Rev 28:475–492
Spönemann J, Brunotte E (1989) Zur Reliefgeschichte der südwestafrikanischen Randschwelle Zwischen Huab und Kuiseb. Z Geomorph Suppl 74:111–125
Spreitzer H (1965) Beobachtungen zur Geomorphologie der Zentralen Namib und ihrer Randgebiete. S.W.A. Sci Soc Special Publication 4:3–34
Stengel HW (1966) The rivers of the Namib and their discharge into the Atlantic: part II: Omaruru and Ugab. Scientific Papers of the Namib Desert Research Station 30
Stengel I (2000) Der Einfluss spätproterozoischer Verwitterung auf die post-Gondwana-zeitliche Reliefgenese im Rooirand-Bereich zwischen Helmeringhausen und Bethanien (Südnamibia). Zbl Geol Palaont 1:623–640
Stengel I (2001) Fossil landslides in the Schwarzrand, Huib and Huns escarpment areas—a contribution to the Quaternary morphostratigraphy of southern Namibia. Palaeoecol Afr 27:221–238
Stengel I, Busche D (2002) Namibia's oldest inselbergs—exhumed landforms of a Late Proterozoic etchplain. In: Yang X (ed) Desert and alpine environments. China Ocean Press, Beijing, pp 121–143
Stone AEC, Thomas DSG, Viles HA (2010) Late Quaternary palaeohydrological changes in the northern namib Sand Sea: new chronologies using OSL dating of interdigitated Aeolian and water-lain interdune deposits. Palaeogeogr Palaeoclimatol Palaeoecol 288:35–53
Strohbach BJ (2008) Mapping the major catchments of Namibia. Agricola 18:63–73
Strohbach BJ (2013) Vegetation of the Okavango River Valey in Kavang West, Namibia. Biodivers Ecol 6:321–339
Swart R, Marais C (2009) Landshapes. The geomorphology of Namibia. Macmillan Education Namibia, Windhoek
Thomas DSG (1984) Ancient ergs of the former arid zones of Zimbabwe, Zambia and Angola. Trans Inst Br Geogr NS 9:75–88
Thomas DSG, Shaw PA (1991) The Kalahari environment. Cambridge University Press, Cambridge
Twidale CR, Maud RR (2013) Auto-exposure and underprinting in the exhumation of Neoproterozoic (subNama) inselbergs in southeastern Namibia. Z Geomorph 57:61–73
Van der Merwe JH (1983) National Atlas of South West Africa (Namibia). South West African Administration, Windhoek
Van der Wateren FM, Dunai TJ (2001) Lake Neogene passive margin denudation history—cosmogenic isotope measurements from the Central Namib Desert. Global Planet Change 30:271–307
Van Zyl JA (1992) The major landform regions of Namibia. S Afr Geogr 19:76–90
Waldron FW (1900) On the appearance and disappearance of a mud island at Walfish Bay. Trans S Afr Phil Soc 11:185–188
Wanke H, Wanke A (2007) Lithostratigraphy of the Kalahari Group in northeastern Namibia. J Afr Earth Sci 48:314–328
Ward JD (1987) The Cenozoic succession in the Kuiseb Valley, Central Namib Desert. Geol Surv South West Africa/Namibia Memoir 9:124

Washington R, Todd M, Middleton N, Goudie AS (2003) Global dust storm source areas determined by the total ozone monitoring spectrometer and ground observations. Ann Ass Am Geogr 93:297–313

Watson AC (1930) The guano islands of southwestern Africa. Geogr Rev 20:631–641

Watson I, Lemon RR (1985) Geomorphology of a coastal desert: the Namib, South West Africa, Namibia. J Coastal Res 1:329–342

Watts NL (1980) Quaternary pedogenic calcretes from the Kalahari (southern Africa): mineralogy, genesis and diagenesis. Sedimentology 27:661–686

Wellington JH (1955) Southern Africa—a geographic study, vol 1., Physical geographyCambridge University Press, Cambridge

Wellington JH (1967) South West Africa and its human issues. Clarendon Press, Oxford

Wieneke F, Rust U (1973) Variations du niveau marin et phases morpholoclimatiques dans le désert du Namib central, Afrique du Sud-Ouest. Finisterra 15:48–65

Wilkinson MJ, Blaha JE, Noli D (1989) A new lagoon on the Namibian coast of South Africa: sand spit growth documented from STS-29 shuttle photography. Geocarto Int 4:63–66

Wilkinson MJ (1990) Palaeoenvironments in the Namib Desert. The lower Tumas basin in the late Cainozoic. University of Chicago Research Papers 231

Zaprowski BJ, Pazzaglia FJ, Evenson EB (2005) Climatic influences on profile concavity and river incision. J Geophys Res 110:F03004. doi:10.1029/2004JF000138

Geology

2

Abstract

The landscapes of Namibia owe much of their distinctive nature to the long and complex geological history. Some of the rocks of Namibia are as much as 2,600 million years old, and the first portion of the chapter gives a chronological picture of the geology of the country. This is followed by an analysis of the importance of Cretaceous tectonic activity associated with the opening of the South Atlantic, and then by a discussion of the significance of dolerite intrusions, of planation surfaces and of current neotectonic activity. All of these geological events have had clear impacts on today's Namibian landscape.

2.1 Geological Background

Namibia has a huge array of different rocks that vary in age and character (Schneider 2008). Their history and nature have been described in a monumental 3-volume treatise by Miller (2008), and a valuable travel handbook has been produced by Grünert (2000). The geological history of Namibia is summarised in the timeline shown in Table 2.1, which focuses on the key events which have shaped today's landscapes. The land that now makes up Namibia has been at the centre of many important plate tectonic movements, making it of great interest to geologists.

A useful distinction can be drawn between the geology of the eastern part of the country and that of the west (Fig. 2.1). The eastern part, comprising the Kalahari basin (see Chap. 1), is geologically young and simple being covered by relatively recent materials, including aeolian sands, alluvium, and calcareous crusts (calcretes). The western part consists of a great variety of rock types exposed in a rugged landscape of valleys, escarpments, mountains and plains and is much more geologically complex and ancient. Many of these rocks were formed at depth in primeval oceans and then were subjected to vertical and horizontal movements of the Earth's crust. The amount of movement, vertical and horizontal, that has been involved is almost incredible. For example, around 500 million years ago, Namibia was located in an area that would now be in the South Pacific and then c 360 million years ago passed over the South Pole, before moving northwards to its present position.

The oldest rocks in Namibia (Early Mokolian to Vaalian) are up to 2,600 million years old and occur in the Hoarusib valley northwest of Sesfontein in the Kunene region of northern Namibia. Between 1,400 and 1,200 million years ago, in the Mesoproterozoic, two or more large landmasses collided to form the great continent of Rodinia (Russian for 'motherland'), producing intense volcanic activity, mountain building and the formation of sedimentary basins. Rodinia started to break up by rifting some 850 million years ago. Africa split from South America, and Namibia was split between the Congo Craton in the north and the Kalahari Craton to the south. Such cratons are stable areas of continental crust that are often referred to as shields. Sediments accumulated in the rifts between these cratons. From 750 million years onwards the three landmasses moved still further apart and deep oceans developed, in which sediments accumulated. However, around 550 million years ago the landmasses fused together once again to form the supercontinent of Gondwanaland. As they converged sediments on the ocean floors were folded and heated and mountains were formed during the Late Precambrian Damaran Orogen. Volcanic rocks were also produced. The rocks of the Damaran Orogenic Belt form a c 400 km wide belt in the centre of Namibia and are well exposed in the Swakop valley (Toé et al. 2013). From 300 to 200 million years ago, rocks of the Karoo Supergroup were deposited, and some of these were laid down by glaciers during the Permo-Carboniferous Dwyka phase. Glacial valleys and pavements were also excavated, and some of them are evident as exhumed forms

A. Goudie and H. Viles, *Landscapes and Landforms of Namibia*, World Geomorphological Landscapes,
DOI 10.1007/978-94-017-8020-9_2,

Table 2.1 A Namibian tectonic and geological timeline

Time (millions of years ago)	Event	Landscape relevance
2,600	Oldest rocks in Namibia formed	Kunene Valley (Chap. 5)
1,600	Start of Mesoproterozoic	
1,400–1,050	Namaqualand metamorphic momplex and related rocks	
850	Break-up of Rodinia	
850–600	Damara Supergroup and Gariep Complex	Otavi limestones and Dolomites (Chap. 6)
650	Damara Granite intrusions	
600–543		
550	Formation of Gondwanaland	
542	End of Precambrian and start of Palaeozoic	
300–180	Karoo Supergroup	
250	Start of Mesozoic	
180	Jurassic. Rifting of Gondwanaland starts from southernmost tip of South America	Etjo sandstones, Waterberg Plateau (Chap. 2, box 3); Twyfelfontein (Chap. 7)
145	Start of Cretaceous	
135–120	Break-up of W Gondwanaland; eruption of Etendeka flood basalts, intrusion of Damaraland igneous complexes and dolerite dikes	Etendeka, Brandberg, Spitzkoppe and Erongo (Chaps. 9, 10 and 11)
80–75	Formation of Bukkaros	Chap. 24
65.5	Start of Cenozoic	
26	Start of Miocene. Formation of Tsondab Sandstone	Calcretes (Chap. 16); Namib Sand Sea Namib sand sea (Chap. 18)
11.8	Intensification of upwelling Benguela current	
7	Start of Pliocene	
7–5	Lake Kunene forms in Etosha Basin	Etosha Pan (Chap. 6)
5–3.7	Roter Kamm formed	Chap. 24
2.5	Start of Quaternary mid-latitude glaciation, and intensification of aridity in Namibia	Homeb Silts (Chap. 15); Sossus Vlei (Chap. 19); Naukluft tufas (Chap. 20); Linear dunes (Chap. 21); Koes Pan (Chap. 23)
0.01	Start of Holocene	Barchans (Chap. 17); Fairy circles (Chap. 25)

in the landscape today (Martin 1968), notably in the upper catchment of the Hoarusib River. The glaciers appear to have been fed by a large ice cap, and the direction of flow was westwards. As the glaciers retreated c 280 million years ago, large spreads of sands and shales were laid down. From 200 to 170 million years ago, an arid phase led to the deposition of sands, some of which now cap the Waterberg Plateau, Mount Etjo, and other hills.

Box 3: The Etjo sandstones: Waterberg Plateau, Mount Etjo and Omatako

Today, a series of flat-topped mountains situated in north eastern Namibia on the edge of the Kalahari and some 60 km east of Otjiwarongo (Fig. 2.2) illustrate the impact of the Etjo Sandstone on Namibia's relief: the Waterberg Plateau (a National Park) (1,930 m above sea level), Klein Waterberg and Mount Etjo (over 2,080 m above sea level). These inselbergs are formed of Karoo age Sandstone, around 180 million years old, underlain by Omingonde shales and mudstones. The Etjo sandstone contains units that were laid down under arid, aeolian conditions (Holzförster et al. 1999; Mountney and Howell 2000). The presence of sandstones over shales and mudstones has produced ideal conditions for slope failures, and great landslips create scallops into the north side of Mount Etjo (Fig. 2.3). Other notable inselbergs are the cone-shaped Omatako Mountains between Okahandja and Otjiwarongo (Fig. 2.4). These were visited by Francis Galton (1853) who likened the perfection of their shapes to that of Tenerife. Their name is based on the Herero word for 'buttock' and they rise 700–800 m

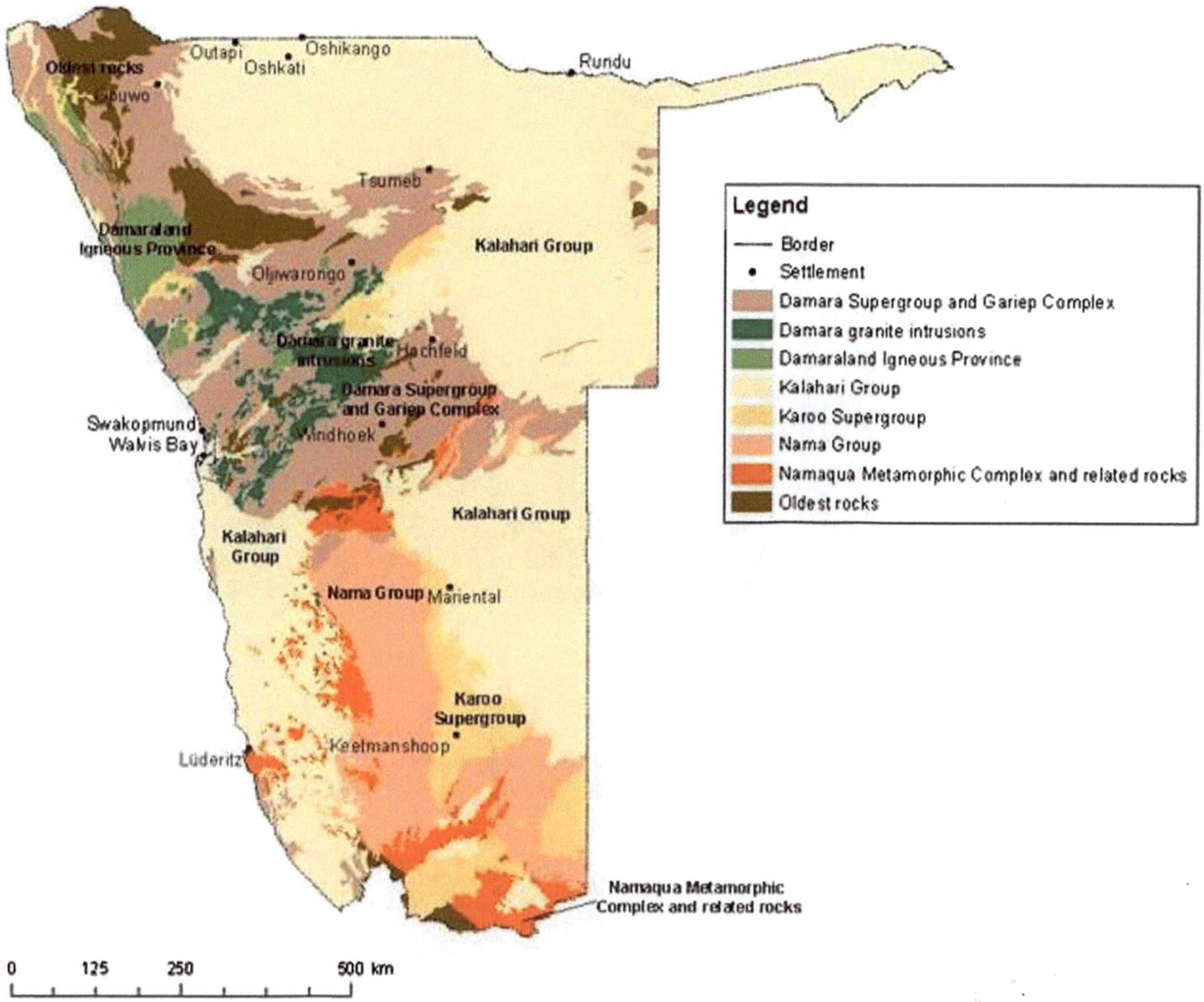

Fig. 2.1 Geological map of Namibia (from Mendelsohn et al. 2002, p. 42, in http://www.uni-koeln.de/sfb389/e/e1/download/atlas_namibia/) (Accessed 30 January 2014)

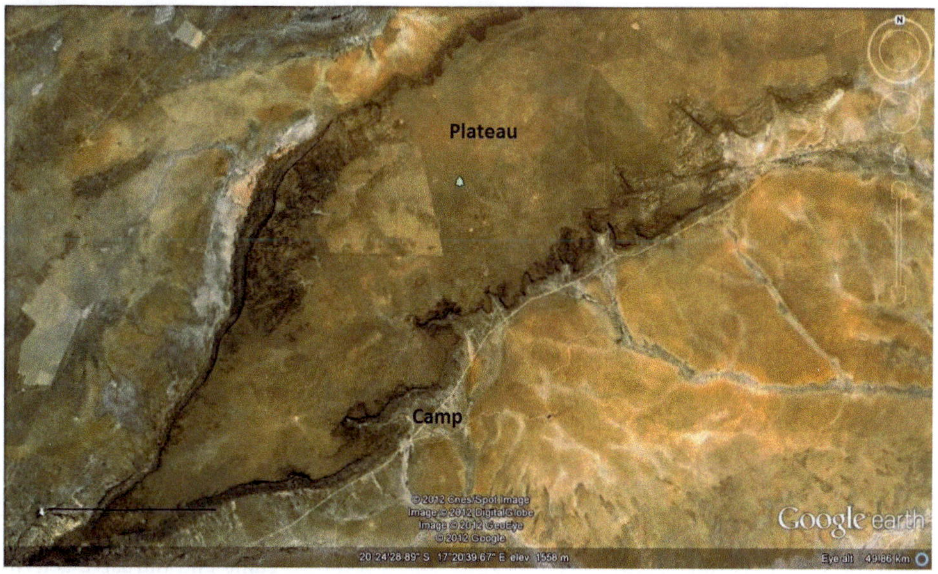

Fig. 2.2 Google Earth image of the Waterberg Plateau. Scale bar is 10 km (© 2012)

Fig. 2.3 Mount Etjo landslides

Fig. 2.4 Google Earth image of Omatako Mountains. Scale bar is 3 km. (© 2013 Digital Globe)

above their surrounding plains. They are also composed of Etjo Sandstone capped by Karoo volcanic rocks (Grünert 2000, p. 33).

Around 180 million years ago, Gondwanaland began to split apart. Great eruptions of lava occurred over southern Africa. The split between South America and South Africa started about 132 million years ago in the early Cretaceous, causing huge eruptions of lava now seen in the Etendeka Mountains as the Etendeka Volcanics (Fig. 2.5). This was a crucial development for understanding the present geomorphology of Namibia (Goudie and Eckardt 1999; Dauteuil et al. 2013), with uplift occurring on the continental margin and subsidence offshore.

2.2 The Impact of Early Cretaceous Tectonics

The Central Namib possesses a group of distinctive igneous complexes—known as Damaraland Complexes—of great relief significance to today's landscape associated with volcanic and intrusive activity caused by the opening up of the

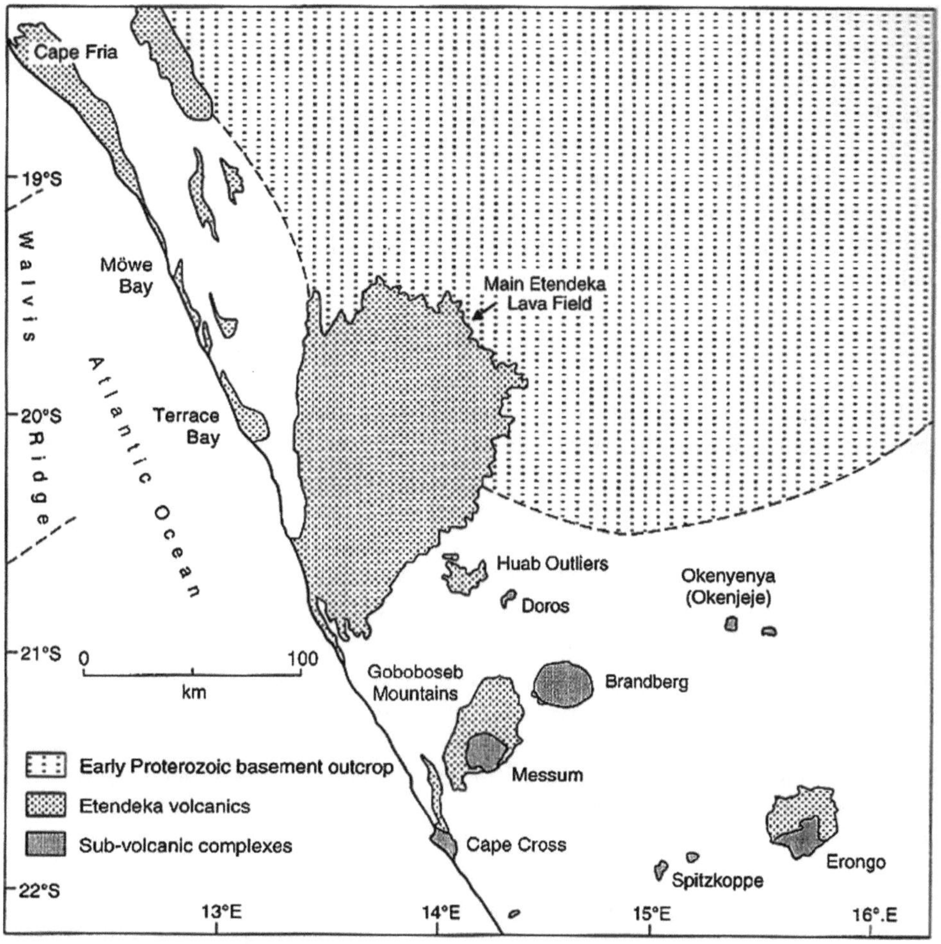

Fig. 2.5 Locations of some of the sub-volcanic complexes and the Etendeka Volcanics (from Goudie and Eckardt 1999, based on Milner et al. 1992, Fig. 1 and Milner and Le Roex 1996, Fig. 1)

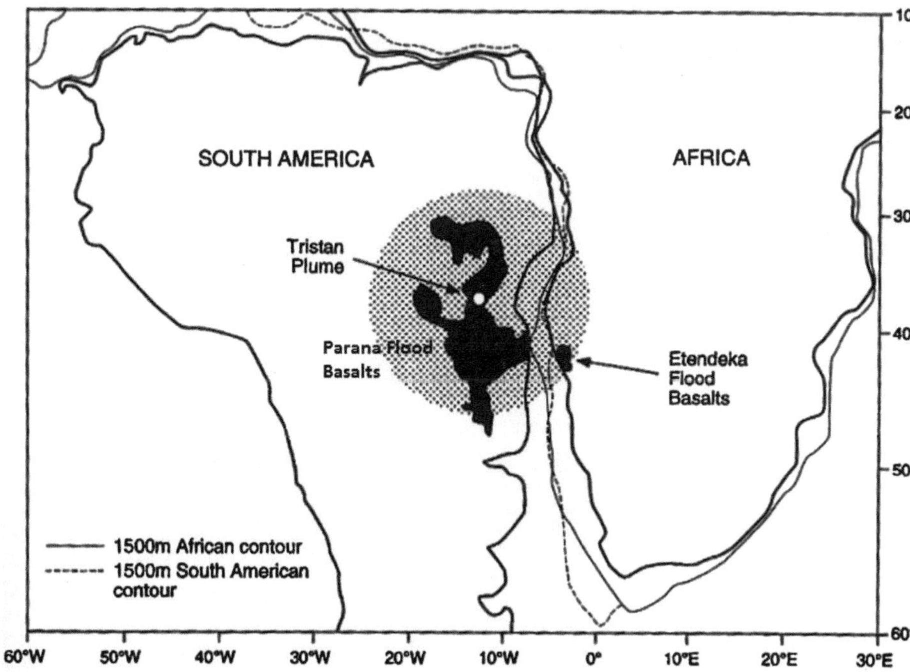

Fig. 2.6 The position of the Etendeka basalts of Namibia in relation to the Paraná basalts in South America at 130 million years ago, prior to the opening of the South Atlantic (from Goudie and Eckardt 1999, Fig. 14 and based on O'Connor and Duncan 1990, Fig. 9a)

South Atlantic, the separation of southern Africa from South America (Fig. 2.6), and the presence of the most important hot-spot track off the coast of Africa—that of Tristan and Gough Islands (Burke 1996), whose topographic expression is the submarine Walvis Ridge (Elliott et al. 2009). Dates for the trail of the Tristan plume along the Walvis Ridge across the South Atlantic show a progressive decrease from Etendeka to the recently active volcanoes of Tristan de Cunha and Gough (O'Connor and Le Roex 1992).

The Damaraland Complexes (Milner et al. 1995) define a prominent northeast trending phenomenon, extending from Cape Cross at the Atlantic coastline, to Okorusu, which is 350 km inland. There are four distinct classes of volcanic/intrusive feature within the area (Pirajno 1994): those formed of granite (Brandberg, Erongo and the Spitzkoppe group), those formed of layered basic rocks (Cape Cross, Doros, Okonjeje, and Messum), those that are peralkaline [e.g. the large Paresis ring complex, located c 280 km inland, which is dominated by rhyolite (Mingram et al. 2000)], and those that are carbonatitic (e.g. Okorusu and Kalkfeld), which are located at the north eastern end of the group. There are also very extensive spreads of continental flood basalts which make up the Etendeka Formation (see Chap. 9) (Fig. 2.7). This outcrops over around 78,000 km^2 of northwestern Namibia and the lavas reach a maximum observed thickness of 800 m at Tafelberg. They are relatively flat-lying and the absence of erosion horizons has been taken to infer that they were poured out rapidly without significant interruption (Milner et al. 1992). Furthermore, numerous dolerite dykes were formed at around 135 Ma, immediately prior to the outpouring of the basalts. They have a strong orientation trend (c NE–SW) and form distinctive ridges. Their trend parallels earlier lineaments that were reactivated during the Damaran deformational event (which started about 850 Ma) and which may be related to still earlier deep-seated Proterozoic crustal weaknesses (Lord et al. 1996).

Of the basic complexes, Cape Cross is a circular feature, the greater part of which lies beneath the seal-rich waters of the south Atlantic. Doros is an elongated pear or funnel shaped mass, with a long axis length of around 7 km. It has been differentially eroded and weathered to give a landscape which consists of alternating ring-shaped hills and valleys, all centred around the middle of the complex. The elevation of each hill or valley lies progressively lower from the centre outwards. A typical annular drainage pattern exists. Okenyenya (Okonjeje) is the best exposed of the differentiated basic complexes and remnants of metamorphosed, outward-dipping Karoo sediments are preserved around its margins. Ring structures dominate the outcrop patterns of the intrusives, which are exposed over an area of about 20 km^2 of high relief. The Messum complex, Brandberg and Spitzkoppe are described in Chap. 10, and Erongo in Chap. 11.

A large number of isotopic dates have now been produced for the Damaraland complexes as reviewed by Milner et al. (1995) which span the 137–124 Ma range and can thus be assigned to the early Cretaceous. Milner et al. (1995) suggest that igneous activity among the different complexes was approximately contemporaneous, and that it was probably concurrent with the eruption of the Etendeka lavas.

2.3 Dolerite Dikes/Dykes and Sills

A dike is a sheet-like intrusion of igneous rock, usually oriented vertically, which cuts across the structural planes of the host rock. If it is more resistant than the rock into which

Fig. 2.7 The Etendeka lavas inland from the Skeleton Coast

it is intruded, it will form an upstanding ridge, whereas if it weathers more than the surrounding rock it will create a ditch-like depression. A sill is a magma body that has been intruded more or less parallel to the bedding of the rocks into which it has been pushed, and so is usually a near horizontal feature in the landscape.

In Namibia there are many dikes composed of dolerite, which rise up above the surrounding land, and because they are often covered in dark desert varnish, they often show up clearly as blackish lines (Fig. 2.8) cutting across the countryside and contrasting with the lighter coloured rocks, such as the granites and marbles around them (Fig. 2.9). Their undulating crests have sometimes been likened to a dragon's backbone. The dikes can occur in swarms, as is the case with the enormous Henties Bay-Outjo NE-trending swarm in west-central Namibia (Trumbull et al. 2004, 2007). This contains well over one thousand individual dikes. They are of roughly the same age as the Etendeka lavas and the intrusive complexes of Damaraland and may be the exposed feeders of the Etendeka basalts.

Another major area of dolerite dikes and sills occurs near Keetmanshoop, but these are older, being of Jurassic (Karoo) age and dating back to 180 million years ago. These were intruded into Dwyka and Ecca sedimentary rocks. These are relatively soft materials so that the dolerites are often left upstanding as small hills. Such is the case at the Giant's Playground, which is located in the Quiver Tree Forest on the Farm Gariganus, 23 km northeast of Keetmanshoop. The dolerites have been weathered to produce rounded hillocks and boulders. The dikes are generally 3–6 m wide, and their length varies from hundreds of metres to tens of kilometres. Mostly they trend NW–SE. One of the first travellers to walk across the Kalahari, Farini (1886), who travelled with his cross-dressing colleague Lulu, thought that the jumble of dikes, boulders and hillocks were man-made and were remnants of a lost city.

2.4 Planation—The African Surface

After the breakup of Gondwanaland in the late Jurassic and early Cretaceous there was in southern Africa through the Cretaceous and into the Miocene a period of comparative stability that led to the erosion of extensive low relief plains, called the African Surface. The surface was possibly eroded under tropical humid conditions, and remnants of kaolinised weathering profiles (saprolite) are locally preserved, not least under Kalahari Group sediments in the Aranos Basin and in the Marienthal-Kalkrand area (Miller 2008, Chap. 23). The surface occurs at low elevations along the coast and at higher levels in the interior behind the Great Escarpment. Above the interior plains some bevelled remnants of the old Gondwana surface remained, as in the rugged Khomas Highlands south of Windhoek (King 1963, p. 240).

The extent of the African Surface in Namibia, and of areas which lie above it, have been mapped in Partridge (1998). In the interior, the African erosion phase is shown by well-developed planation surfaces in the central plateau of the country. This phase was initiated by upwarping of the continental rim associated with rifting and was graded to interior base-levels, such as the Kalahari basin and the Orange valley. In contrast, seaward of the Great Escarpment erosion proceeded to a new, lower, oceanic base level following the rifting. This explains the differences in height of different components of the African Surface. Also, to the seaward of the Great Escarpment, as near Pomona in the Sperrgebiet, the surface was silicified (Partridge and Maud 2000, p.10), as on the Kätchen Plateau (Miller 2008, p. 25–29). The African

Fig. 2.8 Google Earth image of dolerite dykes. Scale bar 1 km. (© 2012 Google Image, GeoEye)

Fig. 2.9 Dolerite dike on the banks of the Swakop River, inland from Swakopmund

erosion phase came to an end with uplift of southern Africa during the Miocene (Moon and Dardis 1988, p. 6).

2.5 Neotectonics

Although Namibia now rests in the middle of a tectonic plate on a passive continental margin, and so has little earthquake activity and no volcanism, there is still probably a certain amount of ongoing tectonic activity in Namibia—a phenomenon called neotectonics. For example, in south west Namibia, the Hebron Fault has a well developed fault scarp, with a downthrow of up to 7 m that can be traced for 35 km, which has disturbed Late Pleistocene dunes (White et al. 2009). In the same area the Dreylingen-Pfalz Fault truncates Pleistocene valley-fill terraces and there are active mud volcanoes offshore of the Orange (Viola et al. 2005). In the northwest the Eisen Graben has had a similar effect on quite recent Kalahari dunes (Wanke 2005). Recent earthquakes in Namibia included a magnitude 5 event near Khorixas on 24th March 2012, and a 5.6 magnitude event on 31st July 2012 between Khorixas and Omaruru. A map of earthquakes in Namibia and neighbouring countries since 1900 appears in: www.earthquake.usgs.gov/earthquakes/world/namibia/seismicity.php (accessed 5th December, 2012).

References

Burke K (1996) The African plate. S Afr J Geol 99:341–409

Dauteuil O, Deschamps F, Bourgeois O, Mocquet A, Guillocheau F (2013) Post-breakup evolution and palaeotopography of the North Namibian Margin during the Meso-Cenozoic. Tectonophysics 589:103–115

Elliott GM, Berndt C, Parson LM (2009) The SW African volcanic rifted margin and the initiation of the Walvis Ridge, South Atlantic. Marine Geophys Res 30:207–214

Farini GA (1886) Through the Kalahari Desert. Sampson Low, London

Galton F (1853) Narrative of an explorer in Tropical South Africa being an account of a visit to Damaraland in 1851. John Murray, London

Goudie AS, Eckardt F (1999) The evolution of the morphological framework of the Central Namib Desert, Namibia, since the early Cretaceous. Geogr Ann 81A:443–458

Grünert N (2000) Namibia. Fascination of geology. Klaus Hess, Windhoek

Holzförster F, Stollhofen H, Stanistreet IG (1999) Lithostratigraphy and depositional environments in the Waterberg-Erongo area, central Namibia, and correlation with the main Karoo Basin, South Africa. J Afr Earth Sci 29:105–123

King LC (1963) South African scenery. Oliver and Boyd, Edinburgh, London

Lord J, Oliver GJH, Soulsby JA (1996) Landsat MSS Imagery of a Lower Cretaceous regional dyke swarm, Damaraland, Namibia: a precursor to the splitting of western Gondwana. Int J Remote Sensing 17:2945–2954

Martin H (1968) Paläomorphologische Formelemente in den Landschaften Südwest-Afrikas. Geol Rund 58:121–128

Mendelsohn J, Jarvis A, Roberts C, Robertson T (2002) Atlas of Namibia. Sunbird Publishers, Cape Town

Miller RMcG (2008) The Geology of Namibia, vol 3. Geological Survey of Namibia, Windhoek

Milner SC, Le Roex AP (1996) Isotope characteristics of the Okenyenya igneous complex, north western Namibia: constraints on the composition of the early Tristan plume and the origins of the EM1 mantle component. Earth Plan Sci Lett 141:277–291

Milner SC, Duncan AR, Ewart A (1992) Quartz latite rheoignimbrite flows of the Etendeka Formation, north-western Namibia. Bull Volcanol 54:200–219

Milner SC, Le Roex AP, O'Connor JM (1995) Age of Mesozoic igneous rock in northwestern Namibia, and their relationship to continental breakup. J Geol Soc London 152:97–104

Mingram B, Trumbull RB, Littman S, Gerstenberger H (2000) A petrographic study of anoregenic felsic magmatism in the Cretaceous Paresis ring complex, Namibia: evidence for mixing of crust and mantle-derived components. Lithos 54:1–22

Moon BP, Dardis GF (1988) The geomorphology of Southern Africa. Southern Book Publishers, Johannesburg

Mountney N, Howell J (2000) Aeolian architecture, bedform climbing and preservation space in the Cretaceous Etjo Formation, NW Namibia. Sedimentology 47:825–849

O'Connor JM, Duncan RA (1990) Evolution of the Walvis Ridge—Rio Grande Rise hot spot system: implications for African and South American plate motions over plumes. J Geophys Res 95 (B11):17475–17502

O'Connor JM, Le Roex AP (1992) South African hot spot—plume systems: 1. Distribution of volcanism in time and space. Earth Plan Sci Lett 13:343–364

Partridge TC (1998) Of diamonds, dinosaurs and diastrophism: 150 million years of landscape evolution in southern Africa. S Afr J Geol 101:167–184

Partridge TC, Maud RR (2000) The Cenozoic of Southern Africa. Oxford University Press, New York

Pirajno F (1994) Mineral resources of anorogenic alkaline complexes in Namibia: a review. Austral J Earth Sci 41:157–168

Schneider G (2008) The roadside geology of Namibia. Gebruder Borntraeger, Berlin and Stuttgart

Toé W, Vanderhaege O, André-Mayer A-S, Feybesse J-L, Milési JP (2013) From migmatites to granites in the Pan-African Damara orogenic belt, Namibia. J Afr Earth Sci 85:62–74

Trumbull RB, Vietor T, Hahne K, Wackerle R, Ledru P (2004) Aeromagnetic mapping and reconnaissance geochemistry of the Early Cretaceous Henties Bay-Outjo dike swarm, Etendeka Igneous Province, Namibia. J Afr Earth Sci 40:17–29

Trumbull RB, Reid DL, de Beer C, van Acken D, Romer RL (2007) Magmatism and continental break up at the west margin of southern Africa: a geochemical comparison of dolerite dikes from northwestern Namibia and the Western Cape. S Afr J Geol 110:477–502

Viola G, Andreoli M, Ben-Avraham Z, Stengel I, Reshef M (2005) Offshore mud volcanoes and onland faulting in southwestern Africa: neotectonic implications and constraints on the regional stress field. Earth Plan Sci Lett 231:147–160

Wanke H (2005) The Namibian Eiseb Graben as an extension of the East African Rift: evidence from Landsat TM 5 imagery. S Afr J Geol 108:541–546

White S, Stollhofen H, Stannistreet IG, Lorenz V (2009) Pleistocene to Recent rejuvenation of the Hebron Fault, SW Namibia. Geol Soc London Special Pub 316:293–317

Climate

3

Abstract

As a result of its location within the global circulation, Namibia is a predominantly dry country, and this chapter describes its main climatic characteristics (rainfall, fog, temperatures, wind etc.) and how they vary across the country. It then investigates the history of climate, examining the possible dates when aridity was established. This is followed by a discussion of climatic changes in the Pleistocene and Holocene.

3.1 Climatic Background

In addition to its geology and tectonic history, the landscapes of Namibia, in line with those of other dry regions (Goudie 2013a) are as they are because of the nature of the country's climate and vegetation, both past and present. Because of its latitudinal position and the presence of the cold offshore Benguela Current, Namibia is mostly dry. Namibia sits within the subtropical high pressure zone at the poleward extent of the tropical Hadley circulation, characterised by descending air and dry surface conditions. Three different circulation patterns exert a particularly strong influence on Namibia's climate. First, the inter-tropical convergence zone to the north moves southwards in summer, allowing moist rain-bearing air into northern Namibia. Second, two anti-cyclonic systems within the sub-tropical high pressure zone are particularly important—the persistent South Atlantic anticyclone offshore which pulls cool air from the south west on to the coast and, especially in winter, the Botswana anticyclone to the east of the country which forces dry air over the country and prevents moist air coming in from the north. Finally, in winter the westerly flow of depressions within the temperate zone to the south moves northwards across southern Africa, allowing cold fronts to penetrate into southern Namibia bringing rain. Under the Köppen system of climatic classification the whole country is classified as B. Within this there are three major types: cool deserts (BWk) along the coast and southwestern interior, warm deserts (BWh) in the southeast and northwest, and semi-desert steppe (BS) in the north and north east. Respectively these cover 17, 36 and 47 % of the country (van der Merwe 1983).

In the coastal strip temperatures are generally modest, and there is a limited seasonal and diurnal range. At the coast temperatures are also strongly oceanic in character, so that they are generally relatively moderate, normally neither being very hot nor very cold. There are low seasonal and daily ranges (c 5 and 8 °C respectively) (Lancaster et al. 1984). Data on average monthly average temperatures for a selection of inland and coastal sites are shown in Table 3.1 and demonstrate the wider seasonal range of temperatures inland (10–12 °C).

Rainfall is sparse in the Namib (Fig. 3.1) and averages c 15 mm per year at Walvis Bay, c 27 mm per year at Gobabeb, and c 90 mm at Ganab (which is c 112 km inland). Towards the base of the Great Escarpment, annual rainfall may exceed 200 mm. There is thus a relatively steady gradient inland. Although the central Namib is hyper-arid, it is not as dry in terms of rainfall as the centre of the Atacama in South America. This is largely a result of the high mountains of the Andes which reinforce aridity in the Atacama and may partially explain why the Atacama has extensive nitrate deposits (caliche) and the Namib does not; as nitrates are extremely soluble and thus likely to become dissolved under even sparse rainfall conditions.

Rainfall in Namibia shows great inter-annual variability (Fig. 3.2). In some years there may be virtually no rain at all, although occasional large events, such as the storms of February 2006 and the April 2006 storm in Lüderitz, which produced 102 mm, do occur, possibly in association with La Niña conditions (Muller et al. 2008). On April 21st 2006 Gobabeb received 24.5 mm in 24 h—around the average annual total in one day. An even wetter day occurred on

A. Goudie and H. Viles, *Landscapes and Landforms of Namibia*, World Geomorphological Landscapes,
DOI 10.1007/978-94-017-8020-9_3, © Springer Science+Business Media Dordrecht 2015

Table 3.1 Mean average temperatures (°C) for selected stations

Month	J	F	M	A	M	J	J	A	S	O	N	D
Coastal												
Elizabeth Bay	17	17	17	16	15	15	14	13	14	14	15	17 (Range 4)
Walvis Bay	17	18	17	16	16	15	14	13	13	13	15	16 (Range 5)
Inland												
Grootfontein	24	23	22	21	17	15	14	17	22	24	25	25 (Range 11)
Rundu	25	25	24	22	19	16	16	19	24	26	26	26 (Range 10)
Windhoek	25	23	22	20	17	13	13	16	20	23	25	25 (Range 12)

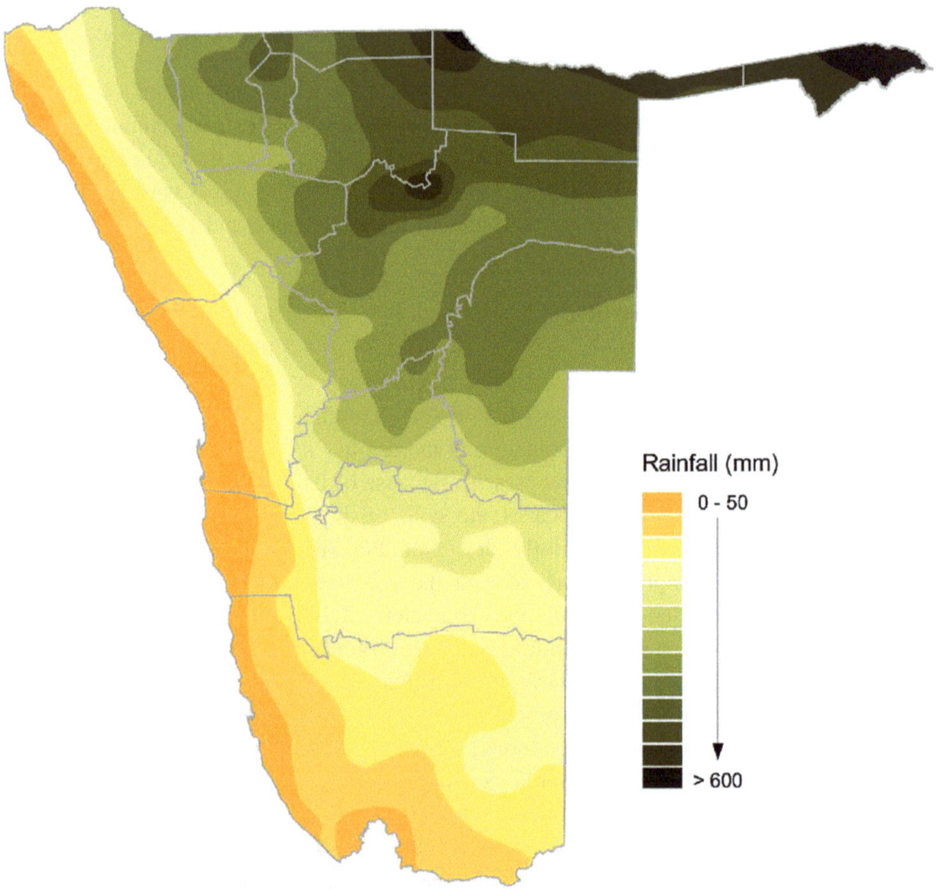

Fig. 3.1 Rainfall map of Namibia (from Mendelsohn et al. 2002, p. 84, in (http://www.uni-koeln.de/sfb389/e/e1/download/atlas_namibia/) (Accessed 30 January, 2014)

March 12, 2011, when 49 mm fell (Eckardt et al. 2012). On March 30, 2013, another storm deposited 38 mm. Such high precipitation events can have great impacts on vegetation dynamics, causing seeds to germinate (Fig. 3.3). Generally, however, individual rainfall events are modest. At Gobabeb in 2001, for example, the annual total of 10.3 mm in that dry year was spread over 12 rainy days, with each rain day experiencing only a few millimeters of precipitation.

The coastal Namib is also a foggy desert with fog on over 100 days in the year in the area around Swakopmund (Fig. 3.4). This precipitates appreciable amounts of moisture. A zone of high fog frequency (>50 fog days per year) "hugs the coast over almost the entire length of the Namib" (Olivier 1995, p. 132), though the highest frequency occurs near Walvis Bay, with 139 fog days per year. Inland from Walvis Bay and Swakopmund the number of fog days is reduced to c 40 within the first 40 km and to c 10 at a distance of 100 km inland. The amount of fog precipitation exceeds rainfall at the coast and mean annual fog precipitation amounts to c 34 mm at Swakopmund, rising to c 183 mm at some of the inland hills such as Swartbank and Vogelfederberg, and then declining to around 3–15 mm further inland (e.g. at Ganab and Zebra Pan) (Lancaster et al. 1984, Table 5).

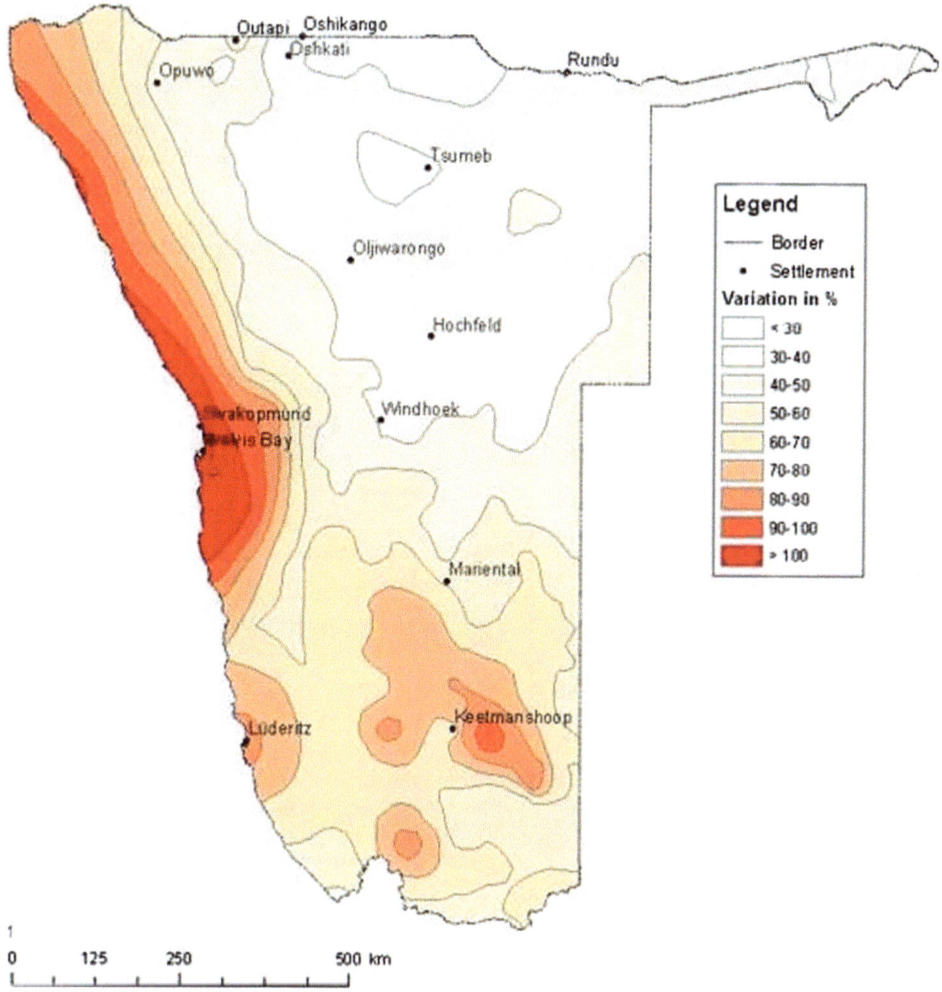

Fig. 3.2 Rainfall variability in Namibia (from Mendelsohn et al. 2002, p. 86, in (http://www.uni-koeln.de/sfb389/e/e1/download/atlas_namibia/) (Accessed 30 January 2014)

Fig. 3.3 In wet years like 2011, grass can sprout densely over wide areas, as here near Gobabeb

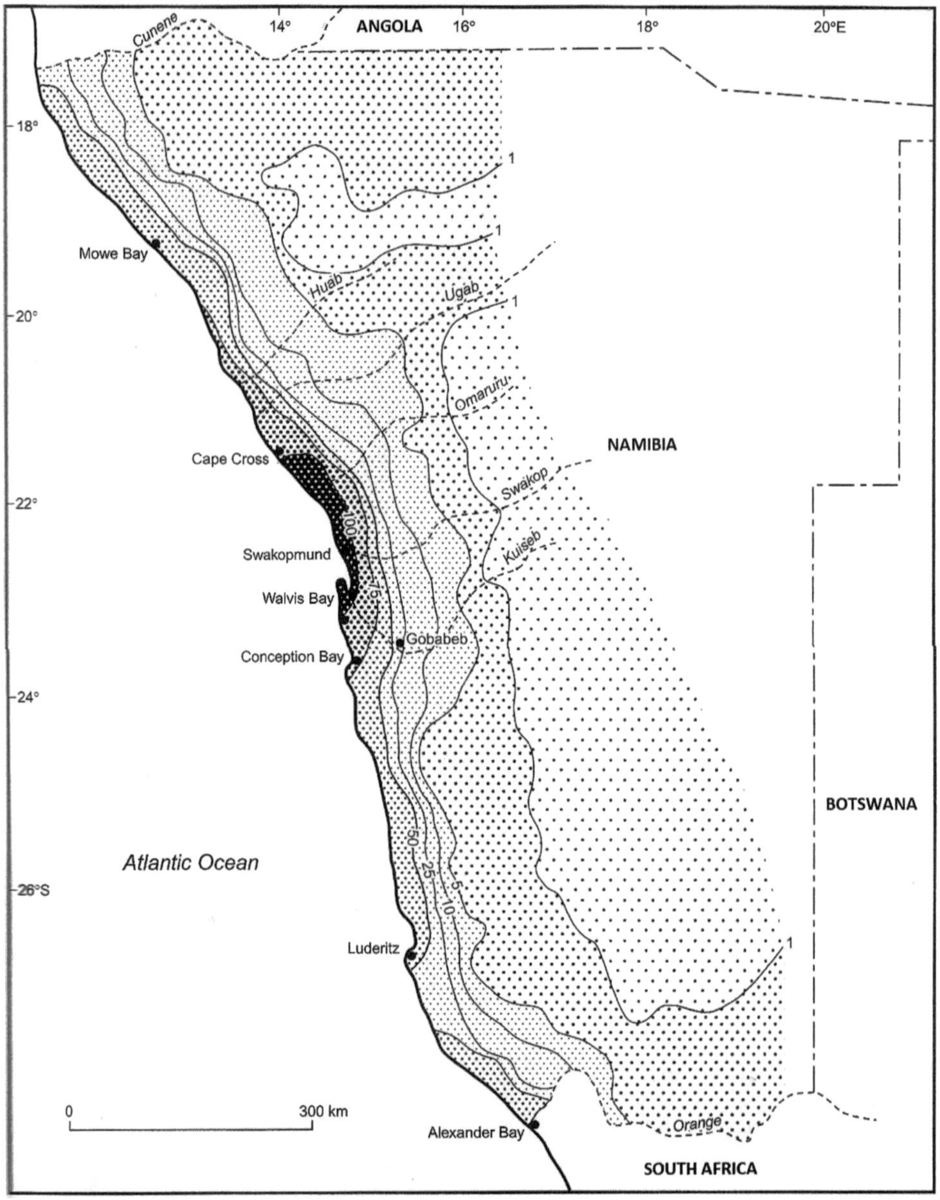

Fig. 3.4 Number of Fog days per year in the Namib (After Olivier 1995, Fig. 1)

Data on daily fog precipitation amounts for three sites along the Kuiseb Valley are provided by Shanyengana et al. (2002) and again demonstrate the considerable quantities of fog that may be deposited on exposed surfaces: 3,308 ml/m^2 at Klipneus, 2,390 at Swartbank, and 508 at Gobabeb (Fig. 3.5). Whilst not all fog days precipitate fog moisture Lancaster et al. (1984, Table 6) indicate the number of fog precipitating days per year as 65 for Swakopmund, 87 for Swartbank, and 3 for Ganab.

In the interior, the climate is much more continental so that summer temperatures can be very high and winter temperatures quite low. Altitude is also an important control of climate, and the central highlands are both slightly cooler and wetter than is the norm. Across the country as a whole, amounts of rain vary from the wettest and most tropical areas in the north east, to the extremely arid Namib Desert in the west. In the north east there are large areas where the mean annual rainfall exceeds 500 mm and in Caprivi it can exceed 600 mm. In the Namib, as we have already seen, the mean annual rainfall is below 50 mm (and in some places as low as 15 mm). Rainfall is highly variable from year to year, especially in the west and south of the country. In general lower falls occur in El Niño years and higher falls during La Niña ones. Over most of the country, there is a summer rainfall regime, though the south west also receives some winter rain. Namibia also suffers from high rates of evapotranspiration, with the highest rates (>2,500 mm per year) occurring in the south east of the country near Keetmanshoop. The whole

Fig. 3.5 Fog at Gobabeb

country potentially loses more water through evaporation than it receives in rain. This is termed a 'water deficit'.

With regard to winds, the most significant winds in the interior come from the east (as shown in Fig. 3.6, Ondangwa and Windhoek), whereas winds from the south and west predominate along the coast (as shown in Fig. 3.6, Lüderitz). This helps to account for the different orientations of dunes found on the coast compared with those further inland (see Chaps. 17 and 18). Sometimes, however, the coast can be subjected to occasionally hot, dry easterly *berg* winds. These often produce dust storms, especially between April and September, which whip up silt from the channels of the Kuiseb delta and the ephemeral streams of the Namib, and then transport it into the Atlantic (Eckardt and Kuring 2005; Tlhalerwa et al. 2005; Vickery and Eckardt 2013; Vickery et al. 2013) (Fig. 3.7). The southern Namib Desert, in the vicinity of Pomona, appears to be especially prone to being swept by high velocity winds, causing wind scouring of bedrock surfaces, yardang formation, and severe deflation. Just inland of Bogenfels major valleys in the Gariep Group schists are aligned roughly north–south, funnelling the southerly winds which pick up in intensity along the valleys. As a result, some of them have been so scoured that their bottoms are now below sea level. In general, wind velocities at the coast are much higher than they are inland. The potential amount of sand which can be moved in a year is called the drift potential and is expressed in vector units. These units are derived from measured wind velocities and their duration, the latter expressed as a percentage of the total measurement period. The longer that high velocity winds blow, the higher the vector units. Annual drift potentials at Walvis Bay are 518 vector units (Fryberger 1979) and at Möwe Bay are 397 vector units (Lancaster 1982). They are highest of all in the southern coastal Namib, where, according to Lancaster (1984) they reach 2,346–2,823 vector units—among the very highest values that have been determined on Earth.

3.2 The Onset of Aridity and an Ancient Namib Desert

Namibia's climate has a long history, and one very important question to raise is just how long-lived the arid climate of the region has been. Like the Damaraland Complexes and the Great Escarpment, the development of aridity in the Namib Desert is closely related to plate tectonics. The existence of arid conditions in the Namib must have been controlled to a considerable extent by three main factors: the opening up of the seaways of the Southern Ocean as the fragments of Gondwanaland moved apart, the location of Antarctica with respect to the South Pole, and the initiation of the offshore, cold Benguela Current (Tankard and Rogers 1978).

The precise date of the initiation of aridity has, however, been a matter of some controversy (Koch 1961; Van Zinderen Bakker 1975; Ward et al. 1983) and Stengel and Busche (2002, p. 122) have gone so far as to argue that 'there has been

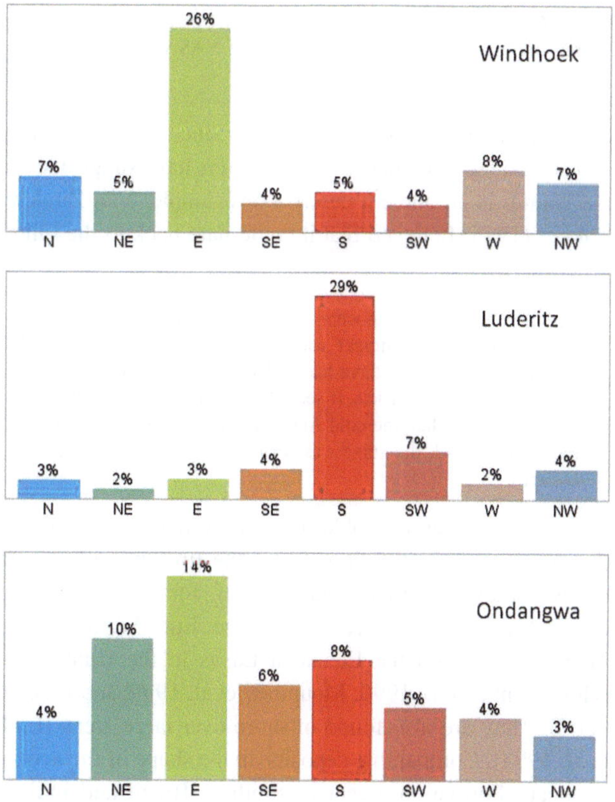

Fig. 3.6 Wind direction frequency (in %) through the year for Windhoek, Lüderitz and Ondangwa (http://weatherspark.com/averages/stations/Namibia) (Accessed 30 January 2014)

Fig. 3.7 MODIS image (June 2004) of dust plumes blowing off the Namib into the Atlantic Ocean (courtesy of NASA)

a largely non-arid landform history since the Late Cretaceous'. On the other hand, many workers have suggested that the Namib is an ancient desert. For example, Van Zinderen Bakker (1975) believed that it could date back to the Oligocene (p. 72):

> … since the Early Oligocene, when the South Atlantic Ocean had developed a sufficient width, its associated climatic and oceanic system could have had a drying influence on the west coast of Southern Africa. It seems therefore that in Early Oligocene times, when the cold Antarctic intermediate water could move northward, the stage was set for the origin of the Namib Desert.

Indeed, it is conceivable that aridity may have been the dominant climatic condition since the time of early Cretaceous continental fragmentation itself, for dune beds (aeolianites) from the upper part of the Etjo Formation are interdigitated with the Etendeka Lavas in the Huab Basin (Horsthemke et al. 1990; Mountney et al. 1999; Jerram et al. 2000). They are also found offshore over large areas (Light et al. 1993). Comparable deposits, in the shape of the aeolian Botucato Formation, occur in southern Brazil and indicate the presence of a large early Cretaceous sand sea prior to the separation of South America and Africa (Scherer 2000). On the other hand, the post-breakup African Surface (see Chap. 1) may have been weathered under moist, tropical conditions (Miller 2008, Chap. 23).

It is probable that the degree of aridity has fluctuated considerably since the early Cretaceous, but that some intensification took place with the establishment of the Benguela Current in the Miocene (Siesser 1980). Support for this scenario was given by Ward et al. (1983) who said (p. 182) that "A review of the Late-Mesozoic-Cenozoic geology leads us to conclude that the Namib tract, which dates back to the Cretaceous, has not experienced climates significantly more humid than semi-arid for any length of time during the last 80 million years". There is the evidence of the Neogene Tsondab Sandstone lithified erg in the southern Namib. This aeolianite dates back to at least the Lower Miocene, and overlies wind-sculptured Late Proterozoic rocks (Senut et al. 1994). The Miocene terrace deposits of the Orange River contain calcretes and gypcretes which formed under relatively dry conditions (Roberts et al. 2013). Likewise, through the study of pollen, charcoal fragments and the stable isotopic composition of plant waxes from an ocean core off Namibia, Hötzel et al. (2013) and Hötzel (2013) suggested that during the Late Miocene there was a great expansion of C_4 grasses (which are particularly adapted to dry conditions).

As already mentioned a crucial control of aridity was the development of the cold Benguela Current and upwelling offshore. Hypotheses to account for the intensification of this current in the last twelve million years include increasing Antarctic glaciation and global cooling, the northward movement of the African continent, the closing of the Central American seaway and African mountain uplift. The most recent studies of the ocean core sediments indicate that the cold upwelling regime developed about 11.8 million years ago (Heinrich et al. 2011) or 10 million years ago (Rommserskirchen et al. 2011), and that soon after that there was an increase in dust inputs to the ocean, pointing to overall drier conditions (Roters and Henrich 2010). Upwelling then continued throughout the entire Pleistocene (Os'Kina and Dmitrenko 2011). The history of cold offshore waters is itself related to the development of ice over Antarctica. Although some glaciation dates back to the latest Eocene (37–34 million years ago), there was a marked intensification of glaciation during the Middle Eocene (c 16–11.6 million years ago) (Lewis et al. 2008; Anderson et al. 2011), and a further expansion in the Late Pliocene (after c 3.3 million years ago) (McKay et al. 2012). The Benguela upwelling system appears to have shown continuous cooling across the Pliocene-Pleistocene from 5 to 3.5 Ma to the end of the Mid-Pleistocene Transition (c 0.6 Ma) (Rosell-Melé et al. 2014). In sum, evidence suggests both that arid conditions were established in the early Cretaceous, and became intensified in the Miocene and Pliocene.

3.3 Quaternary Climatic Change

In most parts of the world, climatic changes in the Quaternary (the last couple of million years) have been crucial in determining landscape character. Such changes were frequent, often abrupt, and very often severe (Anderson et al. 2013). Because of the way in which the world's climatic zones are connected by oceanic and atmospheric circulations, it is inevitable that such events as the multiple great ice ages of the Pleistocene in high and mid latitudes will have had an impact on the lower latitudes. However, both the evidence for climatic change and the reliability of dating of Quaternary sediments in the Namib are relatively poor. There are few depositional basins on shore, and the preservation of pollen is limited. In addition, there have been discrepancies between different dating methods, and problems in interpreting the palaeoenvironmental significance of particular deposits. On the other hand an increasing amount of evidence is now being gained by the study of offshore sediment cores (e.g. Daniau et al. 2013).

Tankard and Rogers (1978) were not convinced of the antiquity of the Namib (p. 334) and believed that while aridity was initiated in the late Tertiary, it only became fully established in the Quaternary. Some evidence of the initiation and maintenance of arid climatic conditions in the Quaternary in Namibia comes from the analysis of offshore sediment cores (e.g. Diester-Haass et al. 1988; Pichevin et al. 2005). For example, Dupont et al. (2005) used pollen analysis to show that rapid desiccation occurred in Namibia at 2.2 Ma (i.e. in the early Pleistocene), and that this was associated with increasing upwelling and decreasing sea-surface temperatures. Maslin et al. (2012), using similar techniques, suggest that there has been very little change in moisture availability over the last 2.5 million years. Lancaster (2002) concurred that the Namib has experienced mostly hyperarid conditions throughout the Quaternary.

However, some investigators have found evidence for relatively wet conditions in the Quaternary. This evidence is diverse. The presence of lake, pond and swamp deposits at scattered localities in the Namib Sand Sea has been used to infer former moister conditions. Lake beds have, for example, been described from Narabeb (23° 41′S, 14° 47′E) (Selby et al. 1979) and from many other locations to the south of the Kuiseb Valley (Teller et al. 1990). Stone et al. (2010) have dated water-lain interdune deposits including those of the Tsondab River, and have found that some of the deposits date back to c 128–75 thousand years (ka) ago (Marine Isotope Stage 5), while others date back less far, to c 16.9–10.5 ka ago. In the Kuiseb Valley and in the Naukluft Mountains there are very extensive tufa deposits, such as the tufas at Blasskrantz in the Naukluft which are around 80 m high and 200 m across (see Chap. 20), which may have formed under wetter conditions in the Quaternary (Ward 1987; Brook et al. 1999). Cave deposits (speleothems) have also proved to be productive of palaeoclimatic information. In addition, some evidence of changes in runoff has been gained from studies of the clay mineral content in offshore marine cores (Gingele 1996).

The results of different investigations are not always consistent in terms of the dating of wet phases in Namibia during the Quaternary and their correlation with other areas. Geyh and Heine (2014), who used ^{230}Th/U dating of a speleothem from Rössing Cave in the Central Namib, identified three relatively wet phases since 420,000 years ago (at 420–385 ka, 230–207 ka, and 120–117 ka). These, they believe, correlated with warm interglacials. Conversely, Stuut et al. (2002) analysed sediment sizes in offshore cores to reconstruct wind velocities and aeolian activity over the last 300,000 years and found that in the Late Quaternary the area was relatively arid during interglacial stages and relatively humid during glacial stages, when the polar front shifted equatorward, resulting in a northward displacement of the zone of westerlies and a consequent increase of rainfall. Shi et al. (2001) have argued that over the last 135 ka there have been six periods during which enhanced south east trade winds contributed to strong upwelling of the Benguela Current and to reduced sea surface temperatures. The most prominent of these occurred c 130 ka BP (Before Present), 42–56 and 17–23 ka BP. They suggest that the good correspondence between the pollen influx record off Namibia and the deuterium isotope record from the Vostok ice core on Antarctica indicates that pronounced glacial Antarctic cooling was accompanied by intensification of the south east trades and enhanced aridity. Heine (1998, p. 190–191) on the other hand concluded that during the last 125,000 ka BP, the hyper-arid coastal zone of the Namib Desert has experienced an arid climate without any precipitation changes greater than those found within the current desert climate.

By contrast, Daniau et al. (2013), using micro-charcoal analysis of an ocean core, found evidence for six fire cycles in the last 170,000 years and related these to changes in rainfall amount and seasonality. An early study of speleothems in a cave at Rössing, inland from Swakopmund, indicated sinter deposition under more humid conditions between 41,500 and 22,500 years BP (Heine and Geyh 1984). Shi et al. (2000) found evidence for three phases of aridification over the past 21,000 years, caused by changes in upwelling and the oceanic thermohaline circulation. Scott et al. (2004) used pollen evidence associated with fossil hyrax dung in the Brandberg massif to show that at the Last Glacial maximum, c 20,000 years ago, the vegetation was dominated by small Asteraceae shrubs, in contrast to the Holocenevegetation which shows more succulents, grass and woody elements. In summary, the Quaternary history of Namibian climate appears complex with much evidence of alternating wet and dry periods, but little agreement over the synchroneity of these with glacial and interglacial conditions elsewhere.

The widespread development of linear dune fields in the Kalahari of southern Angola (Shaw and Goudie 2002), northern Botswana (Grove 1969), and northern and eastern Namibia appears to have taken place at times of greater aridity than today, for many of them are now stable and mantled in vegetation. The Last Glacial Maximum (22–18 ka ago) may have been a time of aridity over most of Namibia (Eitel et al. 2004), with the possible exception of the Caprivi Strip, but Late Quaternary dates for such vegetated ridges in the Caprivi area are provided by Thomas et al. (2000) and Eitel et al. (2004).

With regard to the Holocene, Heine (2005) reviewed a large amount of evidence, and while acknowledging that there may have been greater moisture availability during the early Holocene, and extreme floods during the Little Ice Age c 500 years ago, he argued that overall climatic fluctuations in the Holocene have been modest. However, support for wetter conditions in the early Holocene between c 9 and 5 thousand years ago, is provided by ocean core evidence of great fluvial inputs at that time (Gingele 1996). Similarly, analysis of material preserved in fossil hyrax middens has been productive (Gil-Romera et al. 2006). Fossil hyrax dung from Spitzkoppe has shown relatively wet phases in the Holocene, with four between 8700 and 3500 years ago (Chase et al. 2009). In Kaokoland, analyses of dune sands have indicated more humid conditions at c 8,000 and 4,000 years ago (Bödeker 2006). Aridity is hypothesized to have increased after c 3,800 years ago (Chase et al. 2010). A speleothem from Dante Cave near Tsumeb has also indicated Holocene variability, with a gradual transition from wetter to drier conditions from 4.6 to 3.3 ka BP, a variable but pronounced dry period from 1.8 ka BP to the present with relatively moist conditions between 230 and 100 years BP (AD 1720–1850) (Sletten et al. 2013). Clearly, there still remains much debate about the climatic history of different parts of the Namibian landscape over the Holocene with some evidence of changing levels of aridity, but little agreement over magnitude, timing and causes of such changes.

References

Anderson DE, Goudie AS, Parker AG (2013) Global environments through the Quaternary, 2nd edn. Oxford University Press, Oxford

Anderson JB et al (2011) Progressive Cenozoic cooling and the demise of Antarctica's last refugium. Proc Nat Acad Sci 108:11356–11360

Bödeker O (2006) Sedimentologische und geochronologische Untersuchungen zur jungquartären Klima- und Reliefgeschichte im südlichen Kaokoland (NW-Namibia). Doctoral dissertation, Universität Köln

Brook GA, Marais E, Cowart JB (1999) Evidence of wetter and drier conditions in Namibia from tufas and submerged speleothems. Cimbebasia 15:29–39

Chase BM, Meadows ME, Scott L, Thomas DSG, Marais E, Sealy J, Reimer PJ (2009) A record of Holocene climate change preserved in hyrax middens from southwestern Africa. Geology 37:703–706

Chase BM, Meadows ME, Carr AS, Reimer PJ (2010) Evidence for progressive Holocene aridification in southern Africa recorded in Namibian hyrax middens: implications for African monsoon dynamics and the "African Humid Period". Quat Res 74:36–45

Daniau A-L, Goni MFS, Martinez P, Urrego DH, Bout-Roumazeilles V, Desprat S, Maron HR (2013) Orbital-scale climate forcing of grassland burning in southern Africa. Proc Nat Ac Sci. doi:10.1073/pnas.1214292110

Diester-Haass L, Heine K, Rothe P, Schradre H (1988) Late Quaternary history of continental climate and the Benguela Current off South West Africa. Palaeogeogr Palaeoclim Palaeoecol 65:81–91

Dupont LM, Donner B, Vidal L, Peréz EM, Wefer G (2005) Linking desert evolution and coastal upwelling: Pliocene climate change in Namibia. Geology 33:461–464

Eckardt F, Kuring N (2005) SeaWiFS identifies dust sources in the Namib Desert. Int J Remote Sensing 26:4159–4167

Eckardt FD, Soderberg K, Coop LJ, Muller AA, Vickery KJ, Grandin RD, Jack C, Kapalanga TS, Henschel J (2012) The nature of moisture at Gobabeb, in the central Namib Desert. J Arid Env 93:7–19

Eitel B, Blümel WD, Hüser K (2004) Palaeoenvironmental transitions between 22 ka and 8 ka in monsoonally influenced Namibia. Lect Notes Earth Sci 102:167–194

Fryberger SG (1979) Dune forms and wind regime. US Geol Surv Prof Paper 1052:137–169

Geyh MA, Heine K (2014) Several distinct wet periods since 420 ka in the Namib Desert inferred from U-series dates of speleothems. Quat Res (In Press)

Gil-Romera G, Scott L, Marais E, Brook GA (2006) Middle- to late-Holocene moisture changes in the desert of northwest Namibia derived from fossil hyrax dune pollen. Holocene 16:1073–1084

Gingele FX (1996) Holocene climatic optimum in Southwest Africa—evidence from the marine clay mineral record. Palaeogeogr Palaeoclim Palaeoecol 122:77–87

Goudie AS (2013) Arid and Semi-arid Geomorphology. Cambridge University Press, New York

Grove AT (1969) Landforms and climatic change in the Kalahari and Ngamiland. Geogr J 135:191–212

Heine K (1998) Climate change over the past 135,000 years in the Namib Desert (Namibia) derived from proxy data. Palaeoecol Afr 25:171–198

Heine K (2005) Holocene climate of Namibia: a review based on geoarchives. Afr Study Monographs Suppl 30:119–133

Heine K, Geyh MA (1984) Radiocarbon dating of speleothems from the Rössing cave, Namib Desert, and palaeoclimatic implications. In: Vogel JC (ed) Late Cainozoic palaeoclimates of the Southern Hemisphere. Balkema, Rotterdam, pp 465–470

Heinrich S, Zonneveld KAF, Bickert T, Willems H (2011) The Benguela upwelling related to the Miocene cooling events and the development of the Antarctic Circumpolar Current: evidence from calcareous dinoflagellate cysts. Paleoceanography 26:PA3209, doi:10.1029/2010PA002065

Hötzel S, Dupont L, Schefuss E, Rommerskirchen, Wefer G (2013) The role of fire in Miocene to Pliocene C_4 grassland and ecosystem evolution. Nature Geosci, doi:10.1038/NGEO1984

Hötzel S (2013) Vegetation development in West Africa of the biosphere shift during late Miocene to (early) Pliocene. Dr.rer.nat. dissertation, Faculty of Geosciences University of Bremen

Hortshemke E, Ledendecker S, Porada H (1990) Depositional environments and stratigraphic correlation of the Karoo sequence in north western Damaraland. Comm Geol Surv Namibia 6:63–73

Jerram DA, Mountney NP, Howell JA, Long D, Stollhofen H (2000) Death of a sand sea: an active aeolian erg systematically buried by the Etendeka flood basalts of NW Namibia. J Geol Soc London 157:513–516

Koch C (1961) Some aspects of abundant life in the vegetationless sand of the Namib Desert Dunes. J S West Afr Sci Soc 15:9–92

Lancaster N (1982) Dunes of the Skeleton Coast, Namibia (South West Africa): geomorphology and grain-size relationships. Earth Surf Proc Landf 7:575–587

Lancaster N (1984) Characteristics and occurrence of wind erosion features in the Namib Desert. Earth Surf Proc Landf 9:469–478

Lancaster N (2002) How dry was dry?—Late Pleistocene palaeoclimates in the Namib Desert. Quat Sci Rev 21:769–782

Lancaster J, Lancaster N, Seely MK (1984) Climate of the central Namib desert. Madoqua 14:5–61

Lewis AR, Marchant DR, Ashworth AC, Hedenäs L, Hemming SR, Johnson JV, Leng MJ, Machlus ML, Newton AE, Raine JI, Willenbring JK, Williams M, Wolfe AP (2008) Mid-Miocene cooling and the extinction of tundra in continental Antarctica. Proc Nat Acad Sci 105:10676–10680

Light MPR, Maslanyj MP, Greenwood RJ, Banks NL (1993) Seismic sequence stratigraphy and tectonics offshore Namibia. Geol Soc Special Pub 71:163–191

Maslin MA, Pancost RD, Wilson KE, Lewis J. Trauth MH (2012) Three and a half million year history of moisture availability of South West Africa: evidence from ODP site 1085 biomarker records. Palaeogeogr Palaeoclim Palaeoecol 317–318:41–47

McKay R, Naish, Carter L, Riesselman C, Dunbar R, Sjunneskog C, Winter D, Sangiorgi F, Warren C, Pagani M, Schouten S, Willmott V, Levy R, DeConto R, Powell RD (2012) Antarctic and Southern Ocean influences on Late Pliocene global cooling. Proc Nat Acad Sci: doi:10.1073/pnas.1112248109

Mendelsohn J, Jarvis A, Roberts C, Robertson T (2002) Atlas of Namibia. Sunbird Publishers, Cape Town

Miller RMcG (2008) The geology of Namibia, vol 3. Geological Survey of Namibia, Windhoek

Mountney N, Howell J, Flint S, Jerram D (1999) Climate, sediment supply and tectonics as controls on the deposition and preservation of the aeolian-fluvial Etjo Sandstone Formation, Namibia. J Geol Soc 156:771–777

Muller A, Reason CJC, Fauchereau N (2008) Extreme rainfall in the Namib Desert during late summer 2006 and influences of regional ocean variability. Int J Climatol 28:1061–1070

Olivier J (1995) Spatial distribution of fog in the Namib. J Arid Env 29:129–138

Os'kina NS, Dmitrenko OB (2011) Pleistocene surface water temperatures in the Benguela Upwelling Area. Oceanology 51:652–661

Pichevin L, Cremer M, Giradeau J, Bertrand P (2005) A 190 kyr record of lithogenic grain-size on the Namibian slope: forging a tight link between past wind-strength and coastal upwelling dynamics. Marine Geol 218:81–96

Roberts DL, Sciscio L, Herries AIR, Scott L, Bamford MK, Musekiwa C, Tsikos H (2013) Miocene fluvial systems and palynofloras of the southwestern tip of Africa: implications for regional and global fluctuations in climate and ecosystems. Earth-Sci Rev 124:184–201

Rommerskirchen F, Condon T, Mollenhauer G, Dupont L, Schefuss E (2011) Miocene to Pliocene development of surface and subsurface temperatures in the Benguela Current system. Paleoceanography 26: PA3216. doi: 10.1029/2010PA002074

Rosell-Melé A, Martínez-Garcia A, McClymont EL (2014) Persistent warmth across the Benguela upwelling system during the Pliocene epoch. Earth Plan Sci Lett 386:10–20

Roters B, Henrich R (2010) The middle to late Miocene climatic development of Southwest Africa derived from the sedimentological record of ODP Site 1085A. Int J Earth Sci 99:459–471

Scherer CMS (2000) Eolian dunes of the Botucatu Formation (Cretaceous) in southernmost Brazil: morphology and origin. Sed Geol 137:63–84

Scott L, Marais E, Brook GA (2004) Fossil hyrax dung and evidence of Late Pleistocene and Holocene vegetation types in the Namib Desert. J Quat Sci 19:829–832

Selby MJ, Hendy CH, Seely MK (1979) A late Quaternary lake in the central Namib Desert, southern Africa, and some implications. Palaeogeogr Palaeoclimatol Palaeoecol 26:37–41

Senut B, Pickford M, Ward J (1994) Biostratigraphie de éolianites néogènes du Sud de la Sperrgebeit (Désert de Namib, Namibie). C R Acad Sci 318:1001–1007

Shanyengana ES, Henschel JR, Seely MK, Sanderson RD (2002) Exploring fog as a supplementary water source in Namibia. Atmos Res 64:251–259

Shaw AI, Goudie AS (2002) Geomorphological evidence for the extension of the Mega-Kalahari into south-central Angola. S Afr Geogr J 84:182–194

Shi N, Dupont LM, Beug H-J, Schneider R (2000) Correlation between vegetation in Southwestern Africa and oceanic upwelling in the past 21,000 years. Quat Res 54:72–80

Shi N, Schneider R, Beug H-J, Dupont LM (2001) Southeast trade wind variations during the last 135 kyr: evidence from pollen spectra in eastern South Atlantic sediments. Earth Plan Sci Lett 187:311–321

Siesser WG (1980) Late Miocene origin of the Benguela upwelling system off northern Namibia. Science 208:283–285

Sletten H, Railsback LB, Liang F, Brook GA, Marais E, Hardt BF, Cheng H, Edwards RL (2013) A petrographic and geochemical record of change over the last 4600 years from a northern Namibia stalagmite, with evidence of abruptly wetter climate at the beginning of southern Africa's Iron Age. Palaeogeogr Palaeoclim Palaeoecol 376:149–162

Stengel I, Busche D (2002) Namibia's oldest inselbergs—exhumed landforms of a Late Proterozoic etchplain. In: Yang X (ed) Desert and alpine environments. China Ocean Press, Beijing, pp 121–143

Stone AEC, Viles HA, Thomas L, van Calsteren P (2010) Can ^{234}U-^{230}Th dating be used to date large semi-arid tufas? Challenges from a study in the Naukluft Mountains, Namibia. J Quat Sci 25:1360–1372

Stuut J-BW, Prins MA, Schneider RR, Weltje GJ, Jansen JHF, Postma G (2002) A 300-kyr record of aridity and wind strength in southwestern Africa: inferences from grain-size distributions of sediments on Walvis Ridge, SE Atlantic. Marine Geol 180:221–233

Tankard AJ, Rogers J (1978) Late Cenozoic palaeoenvironments on the west coast of southern Africa. J Biogeogr 5:319–337

Teller JT, Rutter N, Lancaster N (1990) Sedimentology and paleohydrology of Late Quaternary lake deposits in the northern Namib Sand Sea, Namibia. Quat Sci Rev 9:343–364

Thomas DSG, O'Connor PW, Bateman MD, Shaw PA, Stokes S, Nash DJ (2000) Dune activity as a record of late Quaternary aridity in the northern Kalahari: new evidence from northern Namibia interpreted in the context of regional arid and humid chronologies. Palaeogeogr Palaeoclimatol Palaeoecol 156:243–259

Tlhalerwa K, Freiman MT, Piketh SJ (2005) Aerosol deposition off the Southern African West Coast by berg winds. S Afr Geogr J 87:152–161

Van der Merwe JH (ed) (1983) National atlas of South West Africa. Stellenbosch and Windhoek

Van Zinderen Bakker EM (1975) The origin and palaeoenvironment of the Namib Desert biome. J Biogeogr 2:65–73

Vickery KJ, Eckardt FD (2013) Dust emission controls on the lower Kuiseb River valley, Central Namib. Aeolian Res 10:125–133
Vickery KJ, Eckardt FD, Bryant RG (2013) A sub-basin scale dust plume source frequency inventory for southern Africa, 2005–2008. Geophys Res Lett 40:6274–6279
Ward JD (1987) The Cenozoic succession in the Kuiseb Valley. Central Namib Desert, Geol Surv South West Africa/Namibia Memoir 9
Ward JD, Seely MK, Lancaster N (1983) On the antiquity of the Namib. S Afr J Sci 79:175–183

Vegetation, Fauna and Humans

4

Abstract

In response to the variety of climatic conditions that exist, Namibia has a range of vegetation types, with savanna in the north east and deserts in the west and these play an important role in determining the nature and power of geomorphological processes. Wild and domestic animals may also have an important biogeomorphological role. Recently, the human impact has become significant and examples of deliberate and accidental geomorphological consequences of human activities are presented.

4.1 Vegetation and Fauna

One consequence of the present aridity of Namibia is that the vegetation cover is generally low—a closed cover is seldom encountered. A sample of different areas in southern Namibia showed that the average ground cover by vegetation was between 4.7 and 15.3 % (Strohbach 2001), while a more general survey for the whole country (Strohbach et al. 1996) showed the vegetation cover situation for October 1994, just before the main rains. Over 93 % of the country had less than 25 % vegetation cover. The only areas where the vegetation canopy is almost complete is in the woodlands of the north east, including the Caprivi Strip. Whatever the extent and nature of vegetation, it can play many roles in the development of landforms and landscapes for which the umbrella term of biogeomorphology is often used.

A useful measure of the degree of vegetation development in an area is its *biomass,* the total amount of living plant material above and below ground. Deserts in general have a low biomass, often 100 times less than that of an equivalent area of temperate forest. Water is the vital influence on plant growth, of course, and is responsible for this low biomass level. Most plant tissues die if their water content falls too low; the nutrients that feed plants are transmitted by water; water is a raw material in the vital process of photosynthesis; and water regulates the temperature of a plant by its ability to absorb heat and because water vapour lost to the atmosphere during transpiration helps to lower plant temperatures. However, water not only controls the volume of plant matter produced—it also controls their distribution within an area of desert; some areas, because of their soil texture, topographic position or distance from rivers or groundwater, have virtually no water available to plants, whereas others do. The widespread development of banded vegetation (tiger bush) reflects this (see Chap. 25).

In the drier parts of Namibia there are two general classes of vegetation: annuals or *ephemerals,* which have a short lifecycle and may form a fairly dense stand immediately after rain, and *perennials,* which may be succulent and are often dwarfed and woody. The ephemeral plants evade drought. Given a year of favourable precipitation such plants, which include grasses such as *Stipagrostis*, will develop vigorously and produce large numbers of flowers and fruit. This replenishes the seed content of the desert soil. The seeds then lie dormant until the next wet year, when the desert blooms again (see Fig. 3.3).

The perennial vegetation adjusts to the aridity by means of various avoidance mechanisms. Most desert plants are *xerophytes.* They possess drought resisting adaptations: transpiration is reduced by means of dense hairs covering waxy leaf surfaces, by the closure of stomata to reduce transpiration loss and by the rolling up or shedding of leaves at the beginning of the dry season. Some xerophytes, the *succulents*, impound water in their structures. Namibian examples include various types of aloes, euphorbias and commiphoras. Another way of countering drought is to have a limited amount of mass above ground and to have extensive root networks below ground. It is not unusual for the

A. Goudie and H. Viles, *Landscapes and Landforms of Namibia*, World Geomorphological Landscapes,
DOI 10.1007/978-94-017-8020-9_4, © Springer Science+Business Media Dordrecht 2015

Fig. 4.1 Nebkha dunes near Gobabeb

Fig. 4.2 Large herbivores at Okakuejo, Etosha, have probably contributed to pan excavation

roots of some desert perennials to extend downwards more than ten metres. Some plants are woody in type—an adaptation designed to prevent collapse of the plant tissue when water stress produces wilting. Another class of dryland plant is the *phreatophyte*. These have adapted to the environment by the development of long tap roots which penetrate downwards until they approach the assured water supply provided by groundwater. They commonly grow near stream channels, springs or on the margins of lakes and can act as a focus for sand accumulation and nebkha formation (Fig. 4.1). A common phreatophyte in Namibian rivers is the tamarisk (*Tamarix usneoides*).

The pattern of vegetation in Namibia at a gross scale reflects the rainfall pattern (see Chap. 3). Thus plant life is at its lushest and most prolific in the north east and progressively shorter and sparser in the west and south. The dry woodlands of the north-east are in the highest rainfall part of the country (500–700 mm) and merge with the tree savanna of the north-central area. They are characterised by *Baikea plurijugia, Burkea africana, Guibourtia coleosperma* and *Pterocarpus angolensis.* Savanna has variable proportions of trees (e.g. acacias, mopane, etc.), shrubs and grass. Indeed, savannna covers about two thirds of Namibia. The camelthorn savanna (300–400 mm rainfall) of the central Kalahari is an open savanna with *Acacia erioloba* as the dominant tree. Common shrubs include *Acacia hebeclada, Ziziphus mucronata, Tarconanthus camphoratus, Grewia flava, Ozoroa paniculosa* and *Rhus ciliata*. The thornbush savanna (400–500 mm rainfall) is the dominant vegetation type in the central part of the country. Characteristic species include *Acacia reficiens, A. erubescens* and *A. fleckii*. The mopane savanna (50–500 mm rainfall) is a distinct vegetation type dominated by *Colophospermum mopane*, which occurs in tree and shrub forms, in the north-west of the country. In the south of the country is the extensive Nama Karoo biome. This has a varied assemblage of plant communities that range from deciduous shrub vegetation to perennial grasslands and succulent shrubs. In the Namib Desert, except along the ephemeral rivers, vegetation is extremely limited, though in wetter years even the driest parts of the desert can sometimes support a short-lived grass cover. The lack of vegetation in the desert is one reason why there are extensive areas of moving sand and dunes.

The coastal fringe also has extensive lichen fields within the fog belt (Schieferstein and Loris 1992), and these may contribute to geomorphic processes, through enhancing rock weathering and reducing the erosion of desert pavement surfaces. Lichens are able to grow under very dry conditions, and can utilise moisture in fog. Whilst they grow very slowly, and are often small and inconspicuous, they can cover large areas of the land surface. Lichens are effective at trapping dust and protecting fine grained sediment from erosion by wind and runoff.

Namibia, especially in the past, prior to modern-day hunting, was the home of many large mammals, and these have contributed to landscape development through a range of biogeomorphological processes such as trampling (Boelhouwers and Scheepers 2004) and the excavation of pans and river floodplains (e.g. by elephant wallowing) (Ramey et al. 2013) (Fig. 4.2). Animal tracks, produced by both wild and domesticated animals, are an important landscape component in some areas (Fig. 4.3). Smaller organisms, such as termites and ants, have contributed to the formation of various types of patterned ground (see Chap. 25), while in the Walvis Bay lagoon the feeding of flamingos creates round mounds in the coastal muds.

4.2 The Human Impact

Humans have lived in southern Africa for several million years, and the latest episode in the history of the Namibian landscape is their impact on its geomorphology. Humans

Fig. 4.3 Google Earth image of animal tracks in the Otjinjange Valley, northern Namibia. Scale bar 0.10 km (© 2012 Google Image, Google)

modify the landscape in a whole range of ways and their actions have intensified over the past 300 years or so, leading some to champion the term Anthropocene for this latest period of time (Goudie 2013). The extent of human influence on geomorphology over the Anthropocene has been impressive, even in a relatively sparsely populated country like Namibia. For example, humans have excavated large holes in connection with mining activity, as is the case with the uranium mine at Rössing and the diamond mines along the southern coastline (Fig. 4.4). Material excavated in one place may be deposited elsewhere, and waste from the coastal diamond mines affects the intertidal zone and subtidal reefs (Pulfrich et al. 2003) and, as at Elizabeth Bay, can cause beach accretion (Smith et al. 2002). Other deliberate landform modification includes attempts to stop flooding in Walvis Bay by the construction of a flood retention dam across the Kuiseb River, and attempts to stop dune movement and sand encroachment on Walvis Bay's suburbs by erecting sand fences (Le Roux 1974). Humans also affect the landscape accidentally by changing vegetation cover by, for example, the use of fire (Sheuyange et al. 2005), by deforestation (Seely and Klintenberg 2011) and by introducing grazing by domestic stock (Bester 1998/9; Kuiper and Meadows 2002). There are now over 7 million cattle, sheep and goats in Namibia. These factors may cause accelerated soil erosion and rill and gully formation (Strohbach 2000; Eitel et al. 2002) as well as bush encroachment. Desert surfaces, including those covered by lichens, are easily disturbed by off road driving, and this can produce unsightly scars and break up the desert pavement surface, exposing underlying fine materials to wind attack and dust storm generation (Eckardt and White 1997). Other dust storms can

Fig. 4.4 Google Earth image of diamond workings north of the Orange River mouth. Scale bar 1 km (©2012 Terra Metrics, Digital Globe, Google)

Table 4.1 Dams in Namibia with a capacity >5 million m^3

Dam name	River	Date	Capacity (million m^3)
Dreihuk	Hom	1978	15.49
Friedenhau	Kuiseb	1972	6.72
Hardap	Fish	1962	294.59
Naute	Loewen	1972	83.58
Oanab	Oanab	1990	34.51
Olushandja	Kunene	1990	42.33
Omdel	Omaruru	1984	41.29
Omatako	Omatako	1981	43.49
Omatjenne	Omatjenne	1933	5.06
Otivero Main	White Nossob	1984	9.81
Otivero Silt	White Nossob	1984	7.79
Von Bach	Swakop	1970	48.56
Swakoppoort	Swakop	1978	63.49

Source Analysed from data in http://www.namwater.com.na (accessed 13th January, 2013)

Fig. 4.5 Google Earth image of small linear dunes formed in the lee of old mining buildings at Bogenfels Ghost Town. Scale bar 0.10 km (© 2013 GeoEye, Google)

be generated from exposed mine tailings, as at Rosh Pinar (Kříbek et al. 2014). River channels and river sediment loads are being substantially modified by dam and reservoir construction (see Table 4.1) and inter-basin water transfers. For example, the construction of the Von Bach and Swakoppoort dams across the Swakop in the 1970s caused that river's streamflow to be reduced by c 40 % (Marx 2009). Sediment from mining activities can also alter stream channels and their composition (e.g. Taylor and Kesterton 2002). Sand dunes and sand movement can also be affected by human constructions, and this is well illustrated at Bogenfels Ghost Town in the southern Namib, where linear dunes have accumulated in the lee of old buildings (Fig. 4.5). Some of the conservation issues associated with the human impact are discussed in Chap. 26.

References

Bester FV (1998/9) Major problem—bush species and densities in Namibia. Agricola 1–3

Boelhouwers J, Scheepers T (2004) The role of antelope trampling on scarp erosion in a hyper-arid environment, Skeleton Coast, Namibia. J Arid Environ 58:545–557

Eckardt F, White K (1997) Human induced disruption of stone pavement surfaces in the Central Namib Desert, Namibia: observations from Landsat Thematic Mapper. Int J Remote Sens 18:3305–3310

Eitel B, Eberle J, Kuhn R (2002) Holocene environmental change in the Otjiwarongo thornbush savannah (Northern Namibia): evidence from soils and sediments. Catena 47:43–62

Goudie AS (2013) The human impact on the natural environment, 7th edn. Wiley, Oxford

Křibek B, Majer V, Pašava J, Kamona F, Mapani B, Keder J, Ettler V (2014) Contamination of soils with dust fallout from the tailings dam at the Rosh Pinah Area, Namibia: regional assessment, dust dispersion modeling and environmental consequences J Geochem Explor (in press)

Kuiper SM, Meadows ME (2002) Sustainability of livestock farming in the communal lands of southern Namibia. Land Degrad Dev 13:1–15

Le Roux PJ (1974) Drift sand reclamation of Walvis Bay, South West Africa. Int J Biometeorol 18:121–127

Marx VL (2009) Impact of upstream reservoirs on the alluvial aquifer of the Swakop River, Namibia. Diploma thesis, Albert-Ludwigs University, Freiburg

Pulfrich A, Parkins CA, Branch GM, Bustamante RH, Velásquez CR (2003) The effects of sediment deposits from Namibian diamond mines on intertidal and subtidal reefs and rock lobster populations. Aquat Conserv: Mar Freshw Ecosyst 13:257–278

Ramey EM, Ramey RR, Brown LM, Kelley ST (2013) Desert-dwelling African elephants (*Loxodonta africana*) in Namibia dig wells to purify drinking water. Pachyderma 53:66–72

Schieferstein B, Loris K (1992) Ecological investigations on lichen fields in the central Namib 1: distribution patterns and habitat conditions. Vegetatio 98:113–128

Seely MK, Klintenberg P (2011) Case study desertification: Central-northern Namibia. Tropical For 8: doi: 10.1007/978-3-642-19986-8_31

Sheuyange A, Oba G, Weladji RB (2005) Effects of anthropogenic fire history on savanna vegetation in northeastern Namibia. J Environ Manage 75:189–198

Smith G, Mocke G, van Ballegooyen R, Soltau C (2002) Consequences of sediment discharge from dune mining at Elizabeth Bay, Namibia. J Coast Res 18:776–791

Strohbach BJ (2000) Soil erosion—causative factors, extent and prevention. Agro-Info 6:8–14

Strohbach BJ (2001) Vegetation survey of Namibia. J Namibia Sci Soc 49:93–119

Strohbach BJ, Calitz, AJ, Coetzee ME (1996) Erosion hazard mapping: modelling the vegetative cover. Agricola 8:53–59

Taylor MP, Kesterton RG (2002) Heavy metal contamination of an arid river environment: Gruben River, Namibia. Geomorphology 42:311–327

Part II

Regional Studies

5 The North East Kunene Region: The Kunene River, Sand Sea and Yardangs

Abstract

In the far north east of Namibia is a little studied region with a range of interesting landforms, including the valley of the Kunene river itself, which forms the border with Angola, Hartmann Valley, the Kunene erg or sand sea, and extensive areas of wind fluted bedrock (called yardangs). The variety of fluvial forms and dune types is discussed as is the origin of yardangs and their distribution elsewhere in Namibia.

5.1 The Kunene River

The area to the west of the Hartmann Mountains and north of Cape Fria is a remote, wild and largely unstudied area of territory. Within it you can find a huge diversity of landscapes, all produced on ancient Swakop Group rocks, dating back some 850–600 million years and within a hyper-arid climate. The Kunene River has a clearly sinuous course as it cuts across the desert with many waterfalls and rapids. Geology has had a major influence over the river's course, with large faults providing easy paths for it to follow, and resistant bands of rock forming major obstacles which the river has skirted round, or flowed over in the case of waterfalls.

The Kunene River is clearly a potent erosive force. Near Serra Cafema there are a range of dramatic falls cutting into the ancient rocks, and in and around the present channel there are a host of streamlined and sculpted rocks and potholes (Fig. 5.1). Several side channels here probably only experience flows under very high flow conditions, and yet the walls of these channels are gouged with a wide range of large and small potholes. Some of these features are found very high up (sometimes more than 3 m above current flow depths). Whilst hard to date, these erosional features are evidence of formerly high flow conditions (Fig. 5.2).

Another sign of a previously much larger Kunene River in this area is the presence of extensive cobble terraces near Serra Cafema (Fig. 5.3). Brown quartzite cobbles have been deposited in a c. 1 m thick spread over rock surfaces c 5 m above the present floodplain (Nicoll 2010). The rounded quartzite cobbles are not found locally and must have come from many km upstream. The terraces are heavily calcreted, but have not so far been dated in detail.

In contrast, the Hartmann Valley which runs north–south on the west side of the Hartmann Mountains appears to have been formed many millions of years ago during the evolution of the Kunene River system but has had a very different history. It now only contains a very small, ephemeral river system and the entire valley is choked with sediment which has built up over millions of years. On the west side of the Hartmann Valley towards the northern end, are a series of stable and seemingly ancient alluvial fans fronting gneiss and mica schist hills. These understudied alluvial fans are heavily calcreted and covered with gravels.

5.2 The Kunene Sand Sea

The Kunene Sand Sea is located in the far north west of Namibia, and is bounded to the north by the Kunene River, to the west by the Atlantic Ocean, to the south by a large area of wind eroded bedrock in Kaokoland, and to the east by the Hartmann range of mountains (Goudie 2007a). The sand sea is relatively small in size (c 2,000 km^2 in area). Unlike the more southerly Namib Sand Sea (Lancaster 1989) and the Skeleton Coast Sand Sea (Lancaster 1982), its landforms have never been adequately described, except for some dunes in the Hartmann Valley (Hartmann and Brunotte 2008). It is located in a hyper-arid area and it is probable that the mean annual rainfall is less than 50 mm. The nearest climate station in Angola, at Tombua (Porto Alexandre),

A. Goudie and H. Viles, *Landscapes and Landforms of Namibia*, World Geomorphological Landscapes,
DOI 10.1007/978-94-017-8020-9_5, © Springer Science+Business Media Dordrecht 2015

Fig. 5.1 Dramatic falls on the Kunene River to the west of Serra Cafema, looking East with river terraces on south bank

Fig. 5.2 Highly eroded mica schist bedrock sculpted by fluvial erosion near Serra Cafema

Fig. 5.3 Rounded quartzite cobble from Kunene River terrace, near Serra Cafema

generally receives between 10 and 30 mm per annum. Beetz (1934) noted that sand had filled valleys developed in an old land surface (peneplain) cut across ancient rocks and had, in the Cretaceous and Tertiary, entered Angola. He observed sand thicknesses of up to 150 m in sections along the Kunene River.

Analysis of Landsat-7 imagery shows that three main types of dune dominate the Kunene Sand Sea: barchans, transverse and linear. Crescentic barchans, of simple and compound form, cover an area of 348.6 km^2, which is 17.4 % of the total area of the sand sea. They are largely located in the south west of the region and abut the zone of yardang development to their south. Transverse dune ridges are the most extensive dunes of the sand sea, covering c 525.1 km^2 (26.3 % of the area). They dominate the north western part of the region. To the east lies a field of seemingly subdued linear ridges which covers 403.6 km^2 (20.2 % of the area).

The barchans have a mean length of 560 m, and a mean width between their horns of 193 m. Their orientation near the coast is much more north to south than it is further inland. The transverse dune ridges average 2.9 km in length. In general terms dunes in the west and south of the area have northerly orientations, whereas in the north and east of the area they swing round to have an orientation that has a greater west to east component. The Kunene Sand Sea barchans are relatively large compared to the classic examples studied in southern Peru, but they are comparable in width to those from Saudi Arabia, Qatar, Egypt and the southern Namib. Previous studies have shown that rates of barchan movement decrease as dune size increases (e.g. Cooke et al. 1993, Fig. 23.24). These tend to suggest that the Kunene sand sea barchans will move at rates of less than 10 m per year, and probably around c 5 m per year. The linear dunes, which average 7.08 km in length, appear to be overlain by some of the transverse forms, and may, therefore, be of greater age. In the far north east of the sand sea they become increasingly west-east in their orientations. The Curosa-Bahia dos Tigres dune field in Angola, on the other side of the Kunene, shows a similar swinging round of the linear forms as one moves eastwards. It also has transverse dunes overlying linears.

The presence of barchans in the west of the region, in proximity to the ocean, mirrors the situation in the Namib and Skeleton Coast sand seas and may be related to a constancy of wind direction (a necessary prerequisite for barchan development) and also to a limited sand supply. Wind data for stations further south on the north Namibian coast, including Möwe Bay, show strongly unimodal winds coming from the south, whereas inland stations to the east, such as Ondangwa, show more variable wind regimes but with a clear tendency for winds to blow from the east. It is this which probably explains the changing orientations of the dunes as one moves inland.

Fig. 5.4 Landsat 7 image of the yardang and barchan fields to the south of the Kunene River (courtesy of NASA)

5.3 Yardangs

To the south of the Kunene sand sea satellite images indicate that there is a very large area of what appears to be wind-fluted basement rock belonging to the Swakop Group (570–900 Ma). It consists of an expanse of narrow, linear ridges (Fig. 5.4) that trend approximately from south south east to north north west, and appear to have similar orientations in that area to the barchans that move across their surface and to the orientations of the predominant sand streams that have been identified in the Skeleton Coast sand sea to the south. This also corresponds to the predominant wind directions recorded at Mowe Bay, where 62.5 % of winds blow from between 157.5 and 212.5°. The area contains individual ridges running typically for distances of 8–10 km, and with a spacing of around 300–350 m. Field work needs to be done to confirm the precise nature and origin of these features, but it appears possible that they are features called yardangs.

Yardang was introduced by Hedin (1903) as a term for wind abraded ridges of cohesive material. They range in size from small centimetre-scale ridges (micro-yardangs) through to forms that are some metres in height and length (meso-yardangs), to features that may be tens of metres high and some kilometres long (mega-yardangs) (Cooke et al. 1993, pp. 296–297) (Fig. 5.4).

Greeley and Iversen (1985, p. 140) believed that the shape of yardangs, like an upturned ship's hull, was an equilibrium shape, which typically would be 'an elongate hill of 1:4 width-to-length ratio, asymmetric in profile, and with the highest part in the upwind one-third of the hill'. Ward and Greeley (1984) found a 1:4 width to length ratio; Halimov and Fezer (1989) found that the ratios of length, width and height were 10:2:1, while Goudie et al. (1999) found volume, length, width, height ratios of 18.7:9.9:2.7:1.

The forms may go through a cycle of development and eventual obliteration (Halimov and Fezer 1989). Although they are dominantly aeolian erosion features there has been a considerable debate as to the relative importance of deflation, aeolian abrasion, fluvial incision and mass movements in moulding yardang morphology (Goudie 1999). That abrasion is important is indicated by polished, fluted and sand-blasted slopes, and the undercutting of the steep windward face and lateral slopes. It is probably the dominant process in hard bedrock yardangs whereas deflation may be important in the evolution of yardangs developed in soft sediments such as old lake beds. Fluvial erosion may provide an avenue along which wind erosion may occur but excessive fluvial erosion would tend to obliterate yardangs. Mass movements may also be significant when their slopes have been over-steepened by wind erosion.

Goudie (2007b) analysed the key factors that determine the global distribution of mega-yardangs and came up with the following relationships. Firstly, large yardangs occur in hyper-arid areas. Nearly all mega-yardangs occur where rainfall totals are less than 50 mm per annum, whereas pans, as we have seen, become more significant in areas where rainfall is between 150 and 500 mm per year. Large yardangs occur in dry areas where deflation is at a maximum, vegetation cover is minimal, and where sand abrasion can occur.

Secondly, yardangs do not occur in sites of active dune accumulation (e.g. sedimentary basins), though they do occur in former pluvial lake depressions. Basins are areas of sand sea development rather than surface aeolian erosion. Yardangs do not occur in areas with massive alluvial fan accumulation, in truly mountainous areas, or in areas with integrated drainage systems.

Thirdly, mega-yardangs occur in trade wind areas with unidirectional or narrow bimodal wind directions, as is made evident by their association in some cases with barchans (e.g. Northern Namib)—a dune form that only occurs where winds are relatively constant in direction. It is only with such constant wind directions that forms can develop that are parallel to the prevailing wind. They sometimes occur upwind of sand seas, in areas where sand transport occurs (e.g. in Northern and Southern Namibia).

Fourthly, mega-yardangs occur in relatively homogeneous rocks without complex structures (e.g. sandstones), but with jointing along which incision can occur. They do not solely occur in 'soft rocks'.

Yardangs in Namibia are not restricted to the Kunene region. In southern Namibia, between the Namib Sand Sea and the Orange River, there is a hyper-arid area with mega-yardangs developed in ancient crystalline and metamorphic rocks with complex structures (Corbett 1993). Many of the ridges are in excess of 20 km long and are c 1 km across. They run approximately SSE to NNW, which is the same as the trend of individual aeolian dunes in the area. This also corresponds to the dominant annual sand transport directions for Alexander Bay, which lies on the coast to the south of the wind fluted zone (Lancaster 1989, p. 82). Some of the corrasional features near Pomona have previously been recorded as being 100 m high (Krenkel 1928, p. 668). There are at least four main areas where large yardangs occur: just to the south of Lüderitz (c E15° 07′, S 26° 42′); near Pomona (E15° 21′, S 27° 03′; and E 15° 19′, S 27° 09′), and inland from Chamais Bay (E 15° 37′, S 27° 47′).

References

Beetz PFW (1934) Geology of South West Angola, between Cunene and Lunda axis. Trans Geol Soc S Africa 36:137–176

Cooke RU, Warren A, Goudie A (1993) Desert geomorphology. UCL Press, London

Corbett I (1993) The modern and ancient pattern of sandflow through the southern Namib deflation basin. Spec Publ Int Ass Sediment 16:45–60

Goudie AS (2007a) Desert landforms in Namibia – a Landsat interpretation. In: AS Goudie and J Kalvoda (eds). Geomorphological Variations. Nakladatelstvi P3 K, Prague, pp 19–36

Goudie AS (2007b) Mega-yardangs: a global analysis. Geogr Compass 1:65–81

Goudie AS, Stokes S, Cook J, Samieh S, El-Rashidi OA (1999) Yardang landforms from Kharga Oasis, south-western Egypt. Z Geomorphol Suppl 116:1–16

Goudie AS (1999) Wind erosional landforms: yardangs and pans. In Aeolian Environments, Sediments and Landforms, ed. AS Goudie, I Livingstone, S Stokes. Chichester, Wiley, pp 167–180

Greeley R, Iversen JD (1985) Wind as a geological process on Earth, Mars, Venus and Titan. Cambridge University Press, Cambridge

Halimov M, Fezer F (1989) Eight yardang types in Central Asia. Z Geomorphol 33:205–217

Hartmann K, Brunotte E (2008) The genesis and current reshaping of dunes at the eastern end of the northern Namib Desert (Hartmann Valley, NW Namibia). Modelled wind-flow patterns, multi-temporal aerial photograph analysis and anthropogenic morphodynamics. Z Geomorphol Suppl 52:1–14

Hedin S (1903) Central Asia and Tibet. Scribners, New York

Krenkel E (1928) Geologie Afrikas. Reimer, Berlin

Lancaster N (1982) Dunes of the Skeleton Coast, Namibia (South West Africa): geomorphology and grain-size relationships. Earth Surf Proc Landf 7:575–587

Lancaster N (1989) The Namib Sand Sea: dune forms, processes and sediments. Balkema, Rotterdam

Nicoll K (2010) Geomorphic development and Middle Stone Age archaeology of the Lower Cunene River, Namibia-Angola border. Quat Sci Rev 29:1419–1431

Ward AW, Greeley R (1984) Evolution of the yardangs at Rogers Lake, California. Bull Geol Soc Am 95:829–837

6 Etosha Pan and the Karstveld

Abstract

Etosha Pan is one of the iconic landscape features in Namibia, which now acts as a major area of waterholes attracting wildlife. Geomorphologically, it is a large area of internal drainage, which is ephemerally flooded. It has, however, a long history, which has been influenced by both changes in the network of inflowing rivers and changes in climate. Around its shores are old shorelines, lunette dunes, stromatolite accumulations, and spring mounds. The area also has extensive spreads of calcrete. The old limestones and dolomites to the east of Etosha have given rise to a range of karst phenomena, including closed depressions and caves which are also of great scientific and scenic interest.

6.1 Etosha Pan

The Etosha Pan of northern Namibia (Fig. 6.1), which lies at an altitude of 1,070–1,085 m above sea level, is an impressive closed basin which covers 4,760 km^2, with a maximum north to south extent of 80 km and an east to west extent of 120 km. In wet years, when flow comes down the Oshigambo and Ekuma rivers, the Etosha Pan may be flooded across almost its whole surface (Fig. 6.2). This happened in 2011. In dry years (Fig. 6.3) it may dry up almost completely, and then it becomes one of Southern Africa's prime dust storm source areas. In recent years, Etosha has stimulated scientific research interest as a possible analogue for lakes on Titan, Saturn's major moon (Cornet et al. 2012).

Etosha is situated in the southern part of the Owambo basin, which is floored by Mesoproterozoic rocks of the Congo Craton. The basin, which owes its development to the break-up of the Rodinia super-continent, contains some 8,000 m of sedimentary rocks, at the top of which lies the sandy and silty Andoni Formation and the Etosha limestone and calcrete of the Cenozoic Kalahari Group. The Kalahari Beds reach a thickness of as much as 500 m on the Namibia/Angolan border. The Etosha basin is a structural depression (Buch and Trippner 1997) and Etosha Pan is not just the result of wind excavation. The mineralogy and geochemistry of the pan sediments are described by Buch and Rose (1996).

The long-term history of the basin has been discussed by Miller et al. (2010). In the late Miocene, some 5–7 million years ago, the upper Kunene and Okavango rivers fed a large lake, which reached an extent of c 55,000 km^2 at about 3 million years ago. This large lake has been called Lake Kunene (Hipondoka 2005). Subsequently, when a river cutting back from the Atlantic coast captured the headwaters of the Kunene, the lake began to shrink. On its northwest side it was replaced by the shallow, seasonally inundated linear depressions and pools—oshanas—of the Cuvelai system that occur in the north of Namibia. Since the late Pliocene or early Pleistocene, the pan has been modified by water inundation during the rainy season, limestone solution, salt weathering and dust storms (Bryant 2003; Buch 1997; Vickery et al. 2013). Deflation has caused a series of lunette dunes, composed of material derived from the pan floor and from alluvial spreads produced by rivers like the Ekoma, to accumulate as ridges on the western shore (Buch and Zoller 1992; Buch et al. 1992; Hipondoka et al. 2004, 2014). At times, under slightly wetter, low energy, saline conditions, cyanobacteria produced limestones called stromatolites—the Poacher's Point Formation (Smith and Mason 1991; Brook et al. 2013). They have proved difficult to date, but some of them may have formed during a prolonged period of flooding during the early Holocene at c 9–6.6 thousand years ago and others in the Late Pleistocene at c 34–26 thousand years ago (Brook et al. 2011, 2013). The stromatolites are

A. Goudie and H. Viles, *Landscapes and Landforms of Namibia*, World Geomorphological Landscapes,
DOI 10.1007/978-94-017-8020-9_6, © Springer Science+Business Media Dordrecht 2015

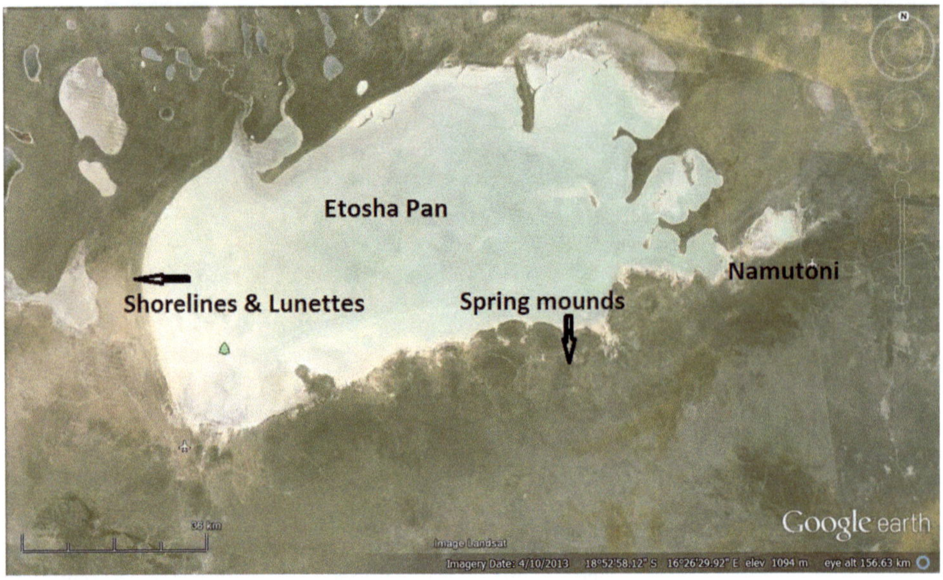

Fig. 6.1 Google Earth image of Etosha Pan. Scale bar is 35 km (© 2013 digital globe)

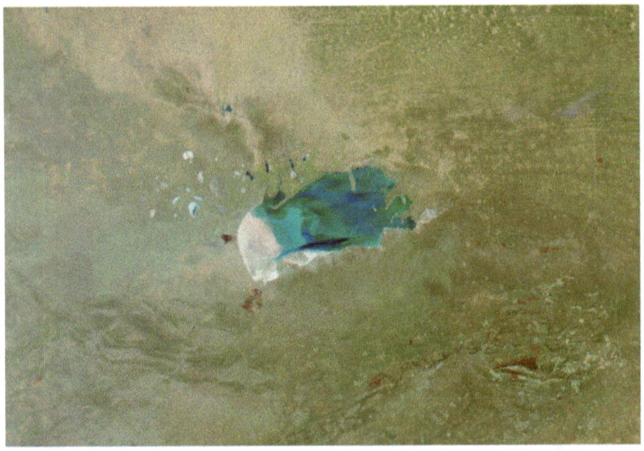

Fig. 6.2 MODIS image of Etosha in July 2011 following heavy rains (courtesy of NASA)

exposed on Pelican Point and Andoni Bay. Fossils have also been found of some semi-aquatic animals such as sitatunga, which are indicative of perennial lake or swamp conditions in the area (Pickford et al. 2009; Hipondoka et al. 2006). Such moist phases created a suite of palaeo-shorelines that are evident on the southwestern boundary of the pan (Brook et al. 2007; Hipondoka et al. 2014). There may have been about seven phases of high lake level over the last 150,000 years (Hipondoka 2005).

One intriguing feature of the Etosha basin, visible on its southern shore and showing up clearly on Google Earth images (e.g. Fig. 6.1), is a series of round landscape features notably on the southern margins of the pan. These are believed to be spring mounds created by sedimentation of chemical matter and the trapping of aeolian and fluvial sediment by the damp, vegetated ground in proximity to springs.

Another interesting and related phenomenon in the area is the development of extensive tracts of carbonate-rich crusts called calcrete (see also Chap. 16). These are associated with groundwater and springs along the contact between the soluble Otavi Group dolomites and limestones and the Mulden Group arkoses and phyllites. In places these calcretes exceed 100 m in thickness, and some of them may be dolomite rich (Miller 2011, p. 21). They form an apron c 80 km wide that skirts the Otavi Group carbonates.

6.2 Karst Landforms

Close to Etosha, to the east, is a very different landscape which records the influence of rainfall in sculpting the surface. A suite of karst landforms are found here. Karst is a type of terrain named after the classic area near Trieste on the borders of Italy and Slovenia. The dominant process that produces karst features is the solution of calcareous rocks, such as limestone, dolomite and marble, by rainwater. In the relatively high rainfall environment of the Otavi Highlands (500–600 mm of rain per year according to Mendelsohn et al. 2002), where ancient limestones and dolomites of the Otavi Group are prevalent, there are many classic landforms produced by such solutional processes—karstification—including various types of solutional rills (*karren*), closed depressions (*dolines*), and caves (as at Ghaub) (Schneiderhöhn 1921). Many of the outcrops near Tsumeb and Otavi appear on remotely sensed images to consist of a multitude of

Fig. 6.3 MODIS image of Etosha in 2002 (a drier year). Note the dust pall blowing out on the west side (courtesy of NASA)

small hillocks interspersed with irregular depressions (Fig. 6.4) and may be roughly analogous to one type of karst characteristic of tropical regions, called cockpit karst. Primate (*Otavipithecus namibiensis*) and hyrax remains have been found in cave breccias of Miocene age (Conroy et al. 1996; Rasmussen et al. 1996). Fossil *Homo* remains have been found at Berg Aukas Cave (Grine et al. 1995) and a mummified baboon in Ludwig Cave (Hodgins et al. 2007). One of the largest caves is Dragon's Breath Cave. This has a large underground lake with an area of almost 2 ha and a depth that recent diving expeditions have indicated exceeds 100 m.

Northwest of the mining town of Tsumeb, are two unusual, deep surface lakes, Otjikoto (Fig. 6.5) and Guinas. The former was visited by Charles James Andersson (1857) in the company of Francis Galton. Andersson, who measured, sketched and swam in it, described it (p. 137) 'as the most extraordinary chasm it was ever my fortune to see', and (p. 138) as 'one of the most wonderful of Nature's freaks'. Otjikoto has a diameter of c 102 m and a depth that may be in excess of 75 m. Guinas is rather larger, with a maximum diameter of 140 m and a depth of 153 m. These two lakes have formed in the carbonate rocks of the Neoproterozoic

Fig. 6.4 Google Earth image of hills and hollows in Otavi Group rocks near Tsumeb. Scale bar 1 km (© 2013 Google Image, © 2014 Digital Globe, © 2014 CNES/Astrium)

Fig. 6.5 Otjikoto

Otavi Group and have resulted from their solution. They are sinkholes created by solution and collapse of surface rocks into large sub-surface caverns and appear to be analogous to the *cenotes* of Yucatan, Mexico. Indeed, the term cenote is often applied to them, and they are most certainly not volcanic craters as White (1969) had once proposed.

References

Andersson CJ (1857) Lake Ngami: or explorations and discoveries during four years wanderings in the wilds of south western Africa. Hurst and Blackett, London

Brook GA, Marais E, Srivastava P, Jordan T (2007) Timing of lake-level changes in Etosha Pan, Namibia, since the middle Holocene from OSL ages of relict shorelines in the Okondeka region. Quat Int 175:29–40

Brook GA, Railsback LB, Marais E (2011) Reassessment of carbonate ages by dating carbonate and organic material from an Etosha Pan (Namibia) stromatolite: evidence of humid phases during the last 20 ka. Quat Int 229:24–37

Brook GA, Cherkinsky A, Railsback LB, Marais E, Hipondoka MH (2013) Radiocarbon reservoir effect in the formation of stromatolites in Namibia. Radiocarbon 55:1156–1163

Bryant RG (2003) Monitoring hydrological controls on dust emissions: preliminary observations from Etosha Pan, Namibia. Geogr J 169:131–141

Buch MW (1997) Etosha Pan—the third largest lake in the world? Madoqua 20:49–64

Buch MW, Rose D (1996) Mineralogy and geochemistry of the sediments of the Etosha Pan region in northern Namibia: a reconstruction of the depositional environment. J Afr Earth Sci 22:355–378

Buch MW, Trippner C (1997) Overview of the geological and geomorphological evolution of the Etosha region, northern Namibia. Madoqua 20:65–74

Buch MW, Zöller L (1992) Pedostratigraphy and thermoluminescence-chronology of the western Margin- (Lunette-) dunes of Etosha Pan, northern Namibia. Würzburger Geogr Arb 84:361–384

Buch MW, Rose D, Zoller L (1992) A TL-calibrated pedostratigraphy of the western lunette dunes of Etosha Pan/northern Namibia: palaeoenvironmental implications for the last 140 ka. Palaeoecol Afr 23:129–147

Conroy GC, Senut B, Gommery D, Pickford M, Mein P (1996) Brief communication: new primate remains from the Miocene of Namibia, southern Africa. Am J Phys Anthropol 99:487–492

Cornet T, Bourgeois O, Le Mouélic S, Rodriguez S, Lopez Gonzalez T, Sotin C, Tobie G, Fleurant C, Barnes JW, Brown RH, Baines KH, Buratti BJ, Clark RN, Nicholson PD (2012) Geomorphological significance of Ontario Lacus on Titan: integrated interpretation of Cassini VIMS, ISS and RADAR data and comparison with the Etosha Pan (Namibia). Icarus 218:788–206

Grine FE, Jungers WL, Tobias PV, Parson OM (1995) Fossil Homo femur from Berg Aukas, northern Namibia. Am J Phys Anthropol 97:151–185

Hipondoka MHT (2005) The development and evolution of Etosha Pan, Namibia. Thesis, University of Würzburg

Hipondoka MHT, Busche D, Kempf J (2004) Source of lunette dune sediments: a geomorphic terrain analysis approach in Etosha National Park, Namibia. Erdkunde 58:212–220

Hipondoka MHT, Jousse H, Kempf J, Busche D (2006) Fossil evidence of perennial lake conditions during the Holocene at Etosha Pan, Namibia. S Afr J Sci 102:93–95

Hipondoka MHT, Mauz B, Kempf J, Packman S, Chiverell RC, Bloemendal J (2014) Chronology of sand ridges and the late Quaternary evolution of the Etosha Pan, Namibia. Geomorphology 204:553–563

Hodgins G, Brook GA, Marais E (2007) Bomb-spike dating of a mummified baboon in Ludwig Cave, Namibia. Int J Speleol 36:31–38

Mendelsohn J, Jarvis A, Roberts C, Robertson T (2002) Atlas of Namibia. Sunbird Publishers, Cape Town

Miller RMG (2011) Calcrete. In: Christelis G, Tuckmeier W (eds) Groundwater in Namibia, an explanation of the hydrogeological map, 2nd edn. Department of Water Affairs, Windhoek, pp 20–22

Miller RMG, Pickford M, Senut B (2010) The geology, palaeontology and evolution of the Etosha Pan, Namibia: implications for terminal Kalahari deposition. S Afr J Geol 113:307–334

Pickford M, Senut B, Hipondoka M, Person A, Segalen L, Plet C, Jousse H, Mein P, Guerin C, Morales J, Mourer-Chauvire C (2009) Mio-plio-pleistocene geology and palaeobiology of Etosha Pan, Namibia. Comm Geol Surv Namibia 14:95–139

Rasmussen DT, Pickford M, Mein P, Senut B, Conroy GC (1996) Earliest known procavid hyracoid from the late Miocene of Namibia. J Mammalol 77:745–754

Schneiderhöhn H (1921) Beiträge zur Kenntnis der Erzlagerstätten und der geologischen Verhältnisse des Otaviberglandes, Deutsch Südwestafrika. Abh Senckenbegischen Naturforschenden Gesellschaft, Frankfurt 37:221–318

Smith AM, Mason TR (1991) Pleistocene, multiple-growth, lacustrine oncoids from the Poacher's Point Formation, Etosha Pan, northern Namibia. Sedimentology 38:591–599

Vickery KJ, Eckardt FD, Bryant RG (2013) A sub-basin scale dust plume source frequency inventory for southern Africa, 2005–2008. Geophys Res Lett. doi:10.1002/grl.50968

White JM (1969) The land God made in anger. The Travel Book Club, London

Twyfelfontein and its Desert Varnish

7

Abstract

Twyfelfontein, a World Heritage Site, which lies in a tributary valley of the Huab River in north western Namibia, is located in an area with Jurassic Etjo Sandstones. These rocks are covered in desert varnish and it is into this material that the famous rock art has been engraved. A discussion is provided of the various ways in which such varnish can develop.

Rocks in Namibia are not always what they seem on the surface, for many of them are covered in a thin layer of a substance called 'desert varnish'. If this dark veneer is removed then the underlying rocks often have a completely different colour. At Twyfelfontein, the varnish has been removed by pecking or chiselling by early inhabitants of the area to produce spectacular rock art, which is since 2007 the basis of a World Heritage Site. This site provides a really good example of the importance of geomorphological processes in influencing cultural heritage. Without the desert varnish the carved rock art would not have been so dramatic. The desert varnish also helps to date the rock art, as it develops within the carved areas over time. The lighter the carved surface the younger the rock art. At Twyfelfontein there are notable representations of many animals, especially giraffe, rhino and zebra (Viereck and Rudner 1957; Vinnicombe 1972) (Fig. 7.1). Other petroglyphs produced by engraving into varnish occur on the granite hills at Piet Albert's near Kamanjab.

Twyfelfontein,/Ui-//aes, located in the Kunene Region of north western Namibia, lies in a tributary valley of the Huab River (Fig. 7.2). The valley is bounded by sandstones of the Jurassic Etjo Formation and shales of the Gai-As Formation, which were laid down in a large lake. These in turn are underlain by dark Kuiseb Formation schists of the Neoproterozoic Damara Sequence. A freshwater spring, thrown out at the junction of the permeable sandstones and the less permeable shales, provided water for early people and also gave rise to the Afrikaans name of the site, for Twyfelfontein means 'doubtful spring'.

The Etjo sandstones, which have an aeolian origin, weather to produce large blocks, rock shelters and clean rock faces. The role of the sandstone as an escarpment former is clearly shown in Fig. 7.2, as is the scatter of large boulders produced by rock falls from the steep face of the escarpment. There are also some magnificent cavernous weathering forms, called tafoni, of which the 'Lion's Mouth' is the most famous example. Other tafoni in Namibia occur on the inselbergs of the Central Namib Plains and their origin is discussed in Chap. 12.

The nature and origin of desert varnish is a matter of some interest. It forms a hard rind, 5 μm to about 100 μm thick (rarely 500 μm), and most varnishes consist of approximately 30 % Mn and Fe oxides and 70 % mixed layer illite/montmorillonite clay minerals (Potter and Rossman 1977). Early workers believed varnishing to be a physicochemical process, associated with high Eh (dry, oxidizing) and pH (unleached, alkaline) conditions. The occasional incoherence of the underlying rock suggested to some that iron and manganese had been drawn in solution from the rock beneath, weakening it internally and then precipitated as a varnish by evaporation. Thus, early models saw varnish constituents as being from the underlying rock (Dorn 1998). Engel and Sharp (1958), while still assuming physicochemical fixation, discovered that many of the elements in varnishes were derived from an external source—dust. It has been

A. Goudie and H. Viles, *Landscapes and Landforms of Namibia*, World Geomorphological Landscapes,
DOI 10.1007/978-94-017-8020-9_7,

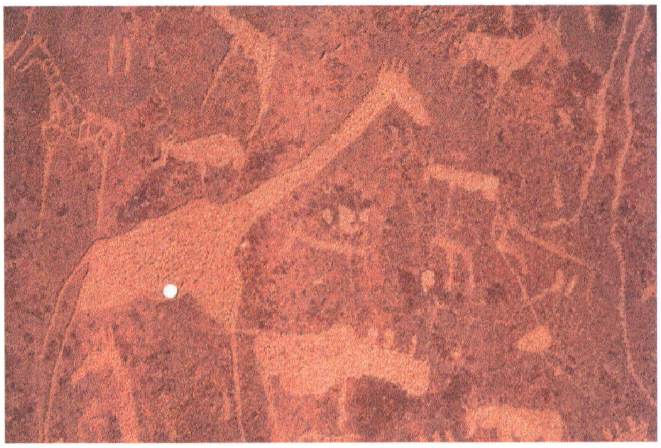

Fig. 7.1 Rock engravings of giraffes and elephant at Twyfelfontein

Fig. 7.2 Google Earth image of the site of Twyfelfontein. Scale bar 350 m. (© 2014 Digital Globe)

demonstrated that most varnish constituents could not have come from underlying rocks, many of which have little iron or manganese. The dominant view is now that varnish is an external accretion (Potter and Rossman 1977), derived not only from dust but also from direct aqueous atmospheric deposition (Thiagarajan and Lee 2004). This view is supported by the presence of radio-isotopes (^{137}Cs and ^{210}Pb) and anthropogenic metals in varnish samples (Fleisher et al. 1999). Moreover, most varnishes overlie the rock with a sharp boundary (Potter and Rossman 1977).

The physicochemical model was challenged by biological models of manganese fixation and this explanation is now widely adopted (Kuhlman et al. 2008). Micro-organisms (lichens, fungi, and bacteria) play a key role in fixing manganese (Dragovich 1993). The increasing use of high resolution geochemical techniques and microscopy has revealed just how complex the composition of varnishes can be and how many processes, biological and physicochemical, contribute to their evolution.

References

Dorn RI (1998) Rock coatings. Elsevier, Amsterdam

Dragovich D (1993) Distribution and chemical composition of microcolonial fungi and rock coatings from arid Australia. Phys Geogr 14:323–341

Engel CE, Sharp RP (1958) Chemical data on desert varnish. Geol Soc Am Bull 69:487–518

Fleisher M, Liu T, Broecker WS, Moore W (1999) A clue regarding the origin of rock varnish. Geophys Res Letters 26:103–106

Kuhlman KR, Venkat P, La Duc MT, Kuhlman GM, McKay CP (2008) Evidence of a microbial community associated with rock varnish at Yungay, Atacama Desert, Chile. J Geophys Res 113:G04022. doi:10.1029/2007JG000677

Potter RM, Rossman GR (1977) Desert varnish: the importance of clay minerals. Science 196:1446–1448

Thiagarajan N, Lee C-TY (2004) Trace-element evidence for the origin of desert varnish by direct aqueous atmospheric deposition. Earth Plan Sci Lett 224:131–141

Viereck A, Rudner J (1957) Twyfelfontein: a centre of prehistoric art in south west Africa. S Afr Arch Bull 12:15–25

Vinnicombe P (1972) Myth, nature, and selection in southern African rock art. Africa 42:192–204

The Ephemeral Rivers and Dunes of the Skeleton Coast

8

Abstract

Along the Skeleton Coast of north western Namibia a series of ephemeral rivers flow into the Atlantic and each demonstrates a very different style of interaction with the area's sand dunes, depending on the width and height of the dune belt they have to traverse. Five different scenarios are described as reflected in the situations found in the Koigab, Uniab, Hunkab, Hoanib and Hoarusib catchments.

The Skeleton Coast on the Atlantic shores of northwest Namibia provides excellent examples of the interrelationship of dunes (aeolian process regimes) and ephemeral rivers (fluvial process regimes). It is crossed by a series of largely east-east flowing major ephemeral rivers (Fig. 8.1). There is also an extensive north to south trending dune belt, composed of barchans and transverse ridges, which starts about 15 km north of the Koigab River. To the south of the Koigab's mouth there is also an area of a dune type that is rare in Namibia, the parabolic or hairpin-shaped dune, which in contrast to a barchan has its nose at the front and trailing ridges behind. There are various different forms of interaction between these rivers and the dunes which gives an array of different landform clusters. Krapf (2003) and Krapf et al. (2003) divide these into a series of types represented by a south to north sequence of catchments. Using space-for-time substitution, also commonly known as ergodic reasoning, they can also be seen as representing different points in a temporal sequence moving from examples where the fluvial regime dominates to those where the aeolian processes dominate, as might happen with progressive desiccation of the climate.

Type 1—the Koigab—occurs where there is no dam created by the dunes and the river is free to reach the ocean. The lower reaches of the Koigab exemplify this well, being characterised by sheet floods and channelized flows in numerous shallow braided channels. It forms a large fan-shaped feature that covers about 120 km^2, and extends over about 10 km from the coastal escarpment towards the Atlantic coastline. The channel extends headwards into the Etendeka Plateau. The fan has been much winnowed by strong wind deflation (Krapf et al. 2009) and the river plays an important role in providing sand for the southern end of the erg.

Type 2—the Uniab—occurs where the river has to pass through a dune belt of limited width and height, which fails to provide a permanent barrier to river flow (Svendsen et al. 2007). The Uniab, which is located near Torra Bay, exemplifies this well, as this river has cut down into the dune belt, forming a broad corridor. The Uniab fan is composed of at least six channels, and as a result of bifurcations eight or nine mouths reach the sea. Fluvial processes dominate and it takes a long time for the healing of aeolian dunes to occur after each flooding event. Past megafloods, accentuated by aeolian damming and then breaching, have deposited organic debris and large boulders of semi-consolidated dune sand, some greater than 10 m in length (Blümel et al. 2000; Svendsen et al. 2003). The river enters the sea via a large waterfall about 1 km east of the present coastline and there is active marine cliffing (Scheepers and Rust 1999) and terrace formation.

A. Goudie and H. Viles, *Landscapes and Landforms of Namibia*, World Geomorphological Landscapes,
DOI 10.1007/978-94-017-8020-9_8, © Springer Science+Business Media Dordrecht 2015

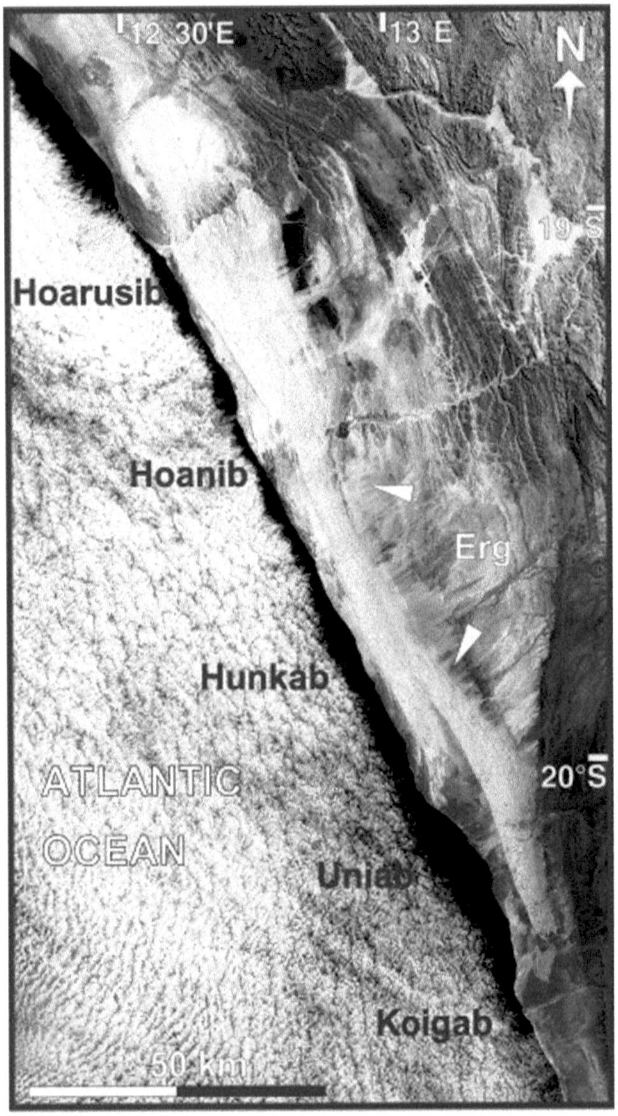

Fig. 8.1 Landsat TM5 image of the Skeleton Coast rivers, 1984 (from Krapf et al. 2003, Fig. 2)

Type 3—the Hunkab—occurs where flow is normally blocked by dunes. This is illustrated by the Hunkab where the river cuts through the erg at times of high flood (as in April, 1995) but then the river bed is covered quickly by dunes after a flood event (Blümel et al. 2000). This river lies between the Uniab and the Hoanib.

Type 4—the Hoanib—occurs where a wide, high dune field stops the river flow (Stannistreet and Stollhofen 2002). Only great floods can force their way through to the sea. This situation is found in the Hoanib catchment in which the extensive 30 × 6 km Gui-Uin floodbasin develops inland from the dunes, with extensive river-end deposits called the Amspoort Silts (Fig. 8.2), some of which are of late Holocene age (Eitel et al. 2005). Large floods are ponded up in interdune corridors and then break through from time to time but healing of the dune belt is very rapid, taking only about 2 years after a flood event. The river only reaches the Atlantic once every 5–10 years (Eitel et al. 2006).

Type 5—the Hoarusib—occurs where the river suppresses the northward advance of windblown sand and successfully reaches the sea most of the time, as is the case for the Hoarusib. Here the river flows regularly and reaches the sea almost every year, enabling large wetlands and dense vegetation to develop along its banks. The Hoarusib itself is notable for the development of 'Clay Castles' caused by aggradation of fine-grained sediments in a tributary immediately upstream of a narrow gorge, c 20 km upstream of the river mouth (Srivastava et al. 2005) (Fig. 8.3). Optical dates suggest deposition in the Late Pleistocene between 44 and 20 ka years ago, under conditions wetter than today. The castles rise up to 70 m above the present river bed.

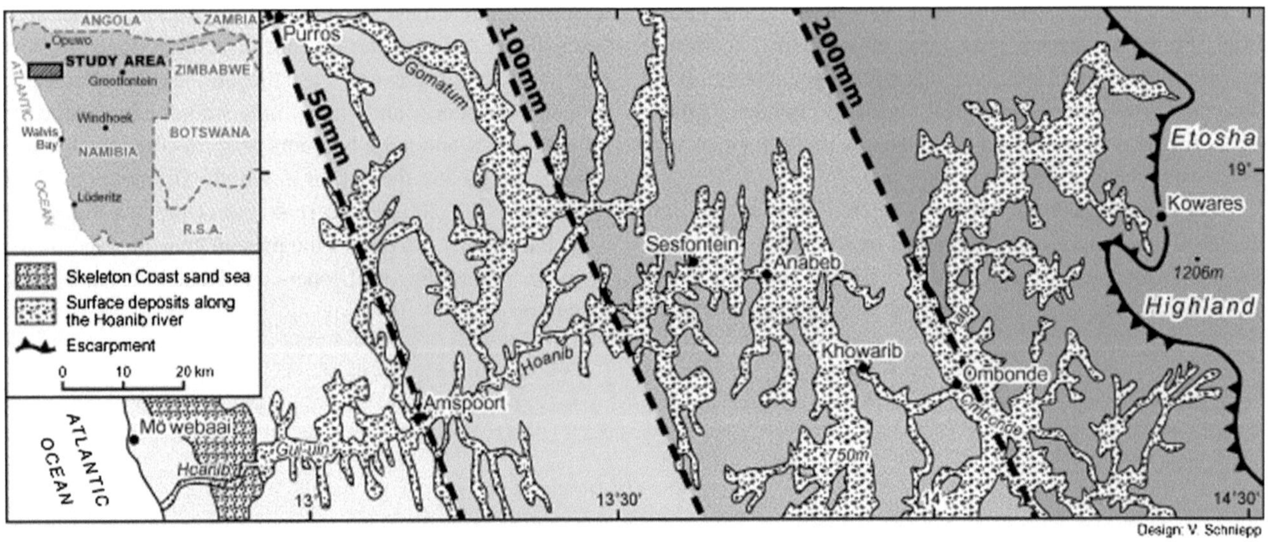

Fig. 8.2 The Hoanib River showing the, showing the extensive surface deposits collectively known as the Amspoort silts (from Eitel et al. 2005, Fig. 1)

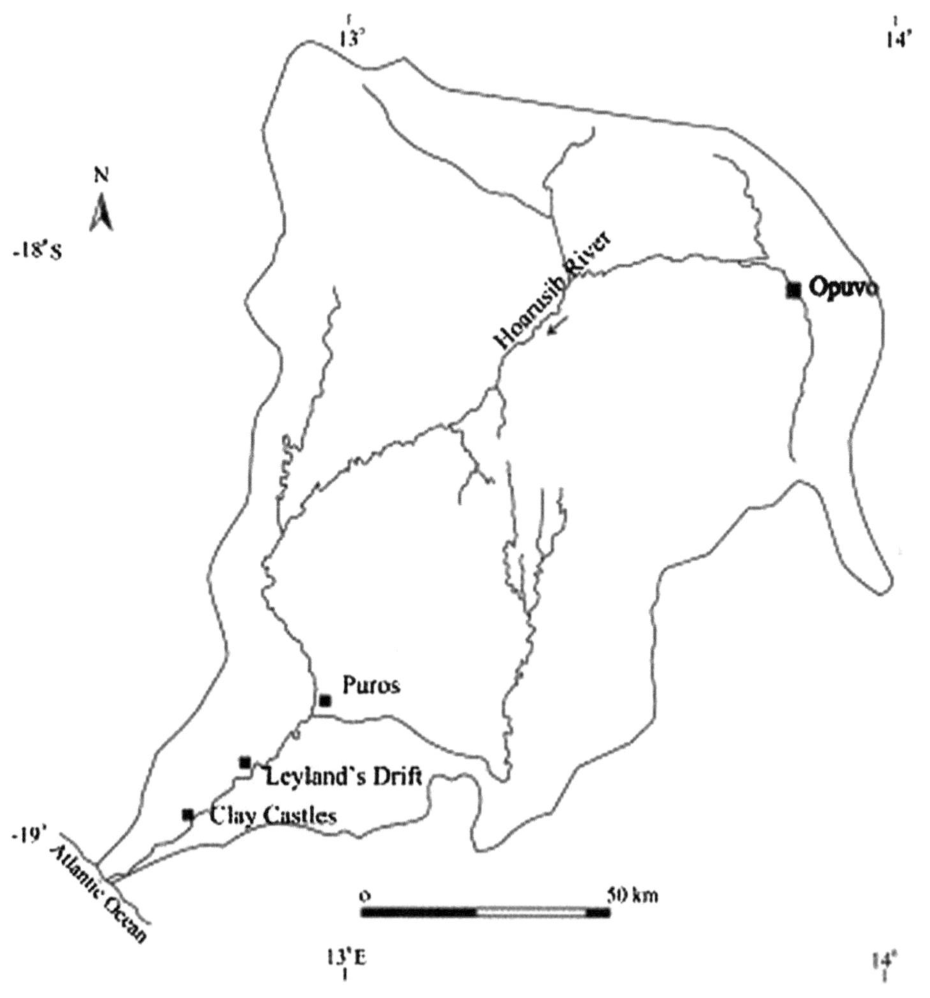

Fig. 8.3 Location of Clay Castles in the Hoarusib basin (from Srivastava et al. 2005, Fig. 1)

References

Blümel WD, Hüser K, Eitel B (2000) Uniab-Schwemmfächer und Skelettküsten-Erg: Zusammenspiel von äolischer und fluvialer Dynamik in der nördlichen Namib. Regensburger Geogr Schrift 33:37–55

Eitel B, Kadereit A, Blümel WD, Hüser K, Kromer B (2005) The Amspoort Silts, northern Namib Desert (Namibia): formation, age and palaeoclimatic evidence of river-end deposits. Geomorphology 64:299–314

Eitel B, Kadereit A, Blümel WD, Hüser, Lomax J, Hilgers A (2006) Environmental changes at the eastern Namib Desert margin before and after the Last Glacial maximum: new evidence from fluvial deposits in the upper Hoanib River catchment, northwestern Namibia. Palaeogeogr Palaeoclim Palaeoecol 234:212–222

Krapf CBE (2003) Ephemeral river systems at the Skeleton Coast, NW-Namibia. Doctoral Dissertation, Universiy of Würzburg

Krapf CBE, Stollhofen H, Stanistreet IG (2003) Contrasting styles of ephemeral river systems and their interaction with dunes of the Skeleton Coast erg (Namibia). Quat Int 104:41–52

Krapf CBE, Stannistreet IG, Stollhofen H (2009) Morphology and fluvio-aeolian interaction of the tropical latitude, ephemeral braided-river dominated Koigab Fan, north-west Namibia. In: Blum MD, Marriott SB, Leclair SE (eds) Fluvial sedimentology VII. Blackwell, Oxford. doi:10.1002/9781444304350.ch6

Scheepers ACT, Rust IC (1999) The Uniab River fan: an unusual alluvial fan on the hyper-arid Skeleton Coast, Namibia. In: Miller AJ, Gupta A (eds) Varieties of fluvial form. Wiley, Chichester, pp 273–294

Srivastava P, Brook GA, Marais E (2005) Depositional environment and luminescence chronology of the Hoarusib River clay castles sediments, northern Namib Desert, Namibia. Catena 59:187–204

Stannistreet IG, Stollhofen H (2002) Hoanib River flood deposits of Namib Desert interdunes as analogues for thin permeability barrier mudstone layers in aeolianite reservoirs. Sedimentology 49:719–736

Svendsen J, Stollhofen H, Krapf CBE, Stannistreet IG (2003) Mass and hyperconcentrated flow deposits record dune damming and catastrophic breakthrough of ephemeral rivers, Skeleton Coast Erg, Namibia. Sed Geol 160:7–31

Svendsen J, Friis H, Stollhofen H, Hartley N (2007) Facies discrimination in a mixed fluvio-eolian setting using elemental whole-rock geochemistry—applications for reservoir characterization. J Sed Res 77:23–33

The Etendeka Plateau

9

Abstract
The Etendeka Plateau is composed of eroded sheets of basaltic lava laid down in the early Cretaceous as a result of sea floor spreading and the opening of the South Atlantic. It is part of one of the world's large igneous provinces and is matched in Brazil by the Paraná basalts. The lavas give rise to striking flat-topped hills and plateaus, which are the eroded remnants of previously more extensive flows.

The Etendeka Plateau of northwestern Namibia is an excellent exemplification of the importance of plate tectonics in landscape evolution (see Chap. 2).

Namibia is now located on what is called a *passive* margin (Gallagher and Brown 1999). That is to say that it is neither in a zone of collision of two or more plates, nor currently at a zone of splitting. It thus has certain similarities to some of the other coastal regions of the fragments of Gondwanaland, such as western India, eastern Brazil or eastern Australia, but is in sharp contract to the *active* margin of western South America where plate collision is creating the very active Andean mountain belt. About 180 million years ago Gondwanaland began to break up into South America-Africa, Australia-Antarctica, and India. The initial opening up of the South Atlantic between South America and Africa began about 135 million years ago.

The Etendeka Plateau of northwestern Namibia, located between the ephemeral Huab River in the south and the Hoanib River in the north, was created by the spewing out of volcanic rocks as the South Atlantic Ocean opened causing an outpouring of flood basalts. These rocks, described in Miller (2008, Chap. 17), have a preserved thickness of 800 m at Tafelberg, and have produced an array of flat-topped hills. The name Etendeka in the local Himba language means 'the place of flat-topped mountains' (Fig. 9.1). The Plateau forms prominent relief and towers 700–800 m above the surrounding plains which are eroded into ancient metamorphic rocks. Layers of different hardness within the lavas produce steps on the sides of the flat-topped towers (Fig. 9.2), and valleys are deeply incised.

The detailed picture which has emerged to explain this vast pile of volcanic material, much of which is basaltic lava, is that in late Jurassic to early Cretaceous times continental break-up along the line of the proto-South Atlantic Ocean was initiated and progressed northwards in a step-like manner. From 150 Ma rifting started at the latitude of the southernmost tip of South America, but not until about 50 million years later was the final breach made in the northern part of the ocean, between South America and Africa. The Paraná basalts of South America and the Etendeka continental flood basalts appear to have erupted at the same time that the northward propagating South Atlantic Rift reached the latitude of Namibia. Particularly dramatic volcanic activity occurred in Damaraland because of the presence of the Tristan Plume, a plume which appears to have been particularly hot and vigorous (Wilson 1992). In South America the basalts cover around 1.5 million km^2, whereas they only cover 78,000 km^2 in Namibia. There are two main hypotheses for these dramatic differences in coverage. First, the spewing out of the lavas was asymmetric, with more being emplaced on the western side (which is now in South America). Second, part of the difference may be accounted for by severe erosion over the last 50 million years or so under the relatively sparse vegetation cover that occurs in Namibia in comparison with western South America.

A. Goudie and H. Viles, *Landscapes and Landforms of Namibia*, World Geomorphological Landscapes,
DOI 10.1007/978-94-017-8020-9_9, © Springer Science+Business Media Dordrecht 2015

Fig. 9.1 Flat-topped mass of Etendeka Lava

Milner et al. (1995) presented a chronological summary of events which was as follows. First, at c 137–125 Ma ago, igneous activity started, initially with limited activity among the Damaraland Complexes (e.g. Cape Cross). Then at 135–132 Ma ago, there was a voluminous outpouring of the Parana-Etendeka volcanic rocks, contemporaneous with, and in response to, the onset of sea-floor spreading.

Several contrasting genetic models for continental flood volcanic events of the Paraná-Etendeka type have emerged over the last two decades. Most models postulate that mantle plumes are responsible (e.g. Richards et al. 1989), though there is debate about their size and structure (Campbell and Griffiths 1990). Where they differ is as to whether melting of the asthenosphere upon rifting above a pre-existing mantle plume generates flood volcanism or, conversely, whether the rise of a mantle plume causes flood volcanism and initiates lithospheric extension. The view of Renne et al. (1992), based on new $^{40}Ar/^{39}Ar$ data for the Paraná flood volcanics, is that the eruption rate was very high, occurred before the initiation of sea-floor spreading in the South Atlantic and was probably precipitated by uplift and weakening of the lithosphere by the Tristan plume. The White and McKenzie model (1989) suggests that flood volcanics form at a continental margin when the opening of a new ocean basin (in this case the South Atlantic) coincides with a mushroom of hot mantle produced by an ascending plume (the Tristan plume). In this model, the sudden onset of volcanic activity is attributed to the passive upwelling of hot mantle as the lithosphere stretches and thins in response to the opening of the ocean. Without very precise dating it is difficult to establish whether effects of the mantle plume precede or follow rifting. There is also considerable debate as to the lateral extent of the plume, though a diameter of 500–2,000 km is often postulated. Such a plume would have caused doming, uplift and drainage pattern modification (Cox 1989).

High rates of Late Cretaceous denudation which have been postulated by some cosmogenic dating studies (see Chap. 1) may account for the widespread removal of sheets of Etendeka lavas, the previous extent of which may be indicated by the occurrence of basaltic dyke swarms that may have been the feeders of now eroded lavas (Trumbull et al. 2004). Certainly, dramatic erosion must have occurred at some stage after the lava was emplaced to produce the complex terrain in this area, with numerous flat-topped hills and plateau features.

The Etendeka/Paraná is one of a number of what are called 'Large Igneous Provinces', which were characterised by rapid and very extensive eruptions of magmatic material. Other examples include Columbia/Yellowstone in North America, Ethiopia/Yemen on either side of the Red Sea, the Deccan plateau of India, and the Karoo volcanics of South Africa. In all cases the extensive outpourings of igneous rocks have left a very visible imprint on today's landscape.

Fig. 9.2 Hills in the Etendeka lava, showing the role of lava layers of different resistance in producing stepped profiles. Note the Himba settlement in the foreground

References

Campbell IH, Griffiths RW (1990) Implications of mantle plume structure for the evolution of flood basalts. Earth Plan Sci Lett 99:79–83

Cox KG (1989) The role of mantle plumes in the development of continental drainage patterns. Nature 342:973–976

Gallagher K, Brown R (1999) Deundation and uplift at passive margins: the record on the Atalntic Margin of southern Africa. Phil Trans R Soc Lond A 835–859

Miller RMG (2008) The geology of Namibia, vol 3. Geological Survey of Namibia, Windhoek

Milner SC, Le Roex AP, O'Connor JM (1995) Age of Mesozoic igneous rock in northwestern Namibia, and their relationship to continental breakup. J Geol Soc London 152:97–104

Renne PR, Ernesto M, Pacca IG, Coe RS, Glen JM, Prevot M, Perrin M (1992) The age of Parana flood volcanism, rifting of Gondwanaland, and the Jurassic-Cretaceous boundary. Science 258:975–979

Richards MA, Duncan RA, Courtillot VE (1989) Flood basalts and hotspot tracks. Science 246:103–107

Trumbull RB, Vietor T, Hahne K, Wackerle R, Ledru P (2004) Aeromagnetic mapping and reconnaissance geochemistry of the Early Cretaceous Henties Bay-Outjo dike swarm, Etendeka Igneous Province, Namibia. J Afr Earth Sci 40:17–29

White R, McKenzie D (1989) Magmatism at rift zones: the generation of volcanic margins and flood basalts. J Geophys Res 94 (B6):7685–7729

Wilson M (1992) Magmatism and continental rifting during the opening if the South Atlantic Ocean: a consequence of lower Cretaceous super-plume activity. Geol Soc London (Special Pub) 68:241–255

Brandberg, Messum and Spitzkoppe

10

Abstract

Several early Cretaceous igneous complexes have produced very distinctive landscapes in central Namibia and in this chapter three are discussed: Brandberg, Messum and the Spitzkoppe group. Brandberg is composed of a circular mass of granite that was intruded at depth and has now been exposed by erosion. Messum, which was originally the centre of a caldera, is a ring complex composed of such rocks as gabbro, rhyolite and syenite. The Spitzkoppe group consists of a series of granitic inselbergs, on which are found striking landforms such as weathering pits, natural arches and rock shelters.

10.1 Brandberg

Three major igneous complexes of early Cretaceous age give some of the most splendid landscapes in Namibia: the Brandberg, Messum and the Spitzkoppe group. Their geology is described in detail by Miller (2008, Chap. 18).

The first of these, the Brandberg, 'the burnt mountain' in German, is called thus because of the colour of its weathered granites (Kirk-Spriggs and Marais 2000). The Damaran name Dâures or Daureb has the same meaning. It is the highest point in Namibia (its Konigstein peak attains almost 2,579 m in altitude). It rises spectacularly above the Late Precambrian schists of the Namib plain (Fig. 10.1) which has a mean elevation of around 700 m above sea level. Located just south of the Ugab River, it has an area of c 420–450 km^2. This makes it is one of the largest sub-volcanic complexes in northwest Namibia. Its nearly circular stock of granite, over 20 km in diameter, and so plainly displayed on satellite images, is surrounded by a collar or skirt of Permo-Triassic Karoo sediments and remnants of basalt and quartz latite of the Etendeka Group. To the north of Brandberg, across the Ugab River, is an area of modest sand dunes.

Brandberg was a volcano at least 25 km in diameter and 2 km high that was active immediately after eruption of the Etendeka volcanic rocks (Miller 2000). Following cauldron subsidence on a ring fracture 25 km in diameter, a sequence of rings and plugs of granite were emplaced into the volcano (Fig. 10.2). This emplacement occurred about 130 million years ago. The thermally metamorphosed Karoo strata generally dip towards the complex, having been dragged downward by subsidence of the granite along the peripheral ring-fault.

The great dome (Fig. 10.3), created by the intrusion of the plutonic material, has since been altered in shape as a result of the erosion of the great bulk of the overlying cover of Karoo rocks (Dauteuil et al. 2013). Apatite fission track analysis (Raab et al. 2005) suggests that there may have been as much as 5 km of denudation in the Brandberg region since the Late Cretaceous. Much of this took place in the Late Cretaceous itself, with rates of 200 m per million years, declining to an average of <20 m per million years in the Tertiary. This tends to confirm the lower rates of current denudation implied by some studies based on long term dating by cosmogenic nuclides (see Chap. 1 for further details).

Brandberg is not only notable for its geomorphology, with its deep gorges cut into masses of granite and granite boulders, but it is also the location of a rich array of archaeological sites, of which the so-called White Lady rock painting in the Tsisab Gorge, is the most famous.

A. Goudie and H. Viles, *Landscapes and Landforms of Namibia*, World Geomorphological Landscapes,
DOI 10.1007/978-94-017-8020-9_10, © Springer Science+Business Media Dordrecht 2015

Fig. 10.1 Brandberg

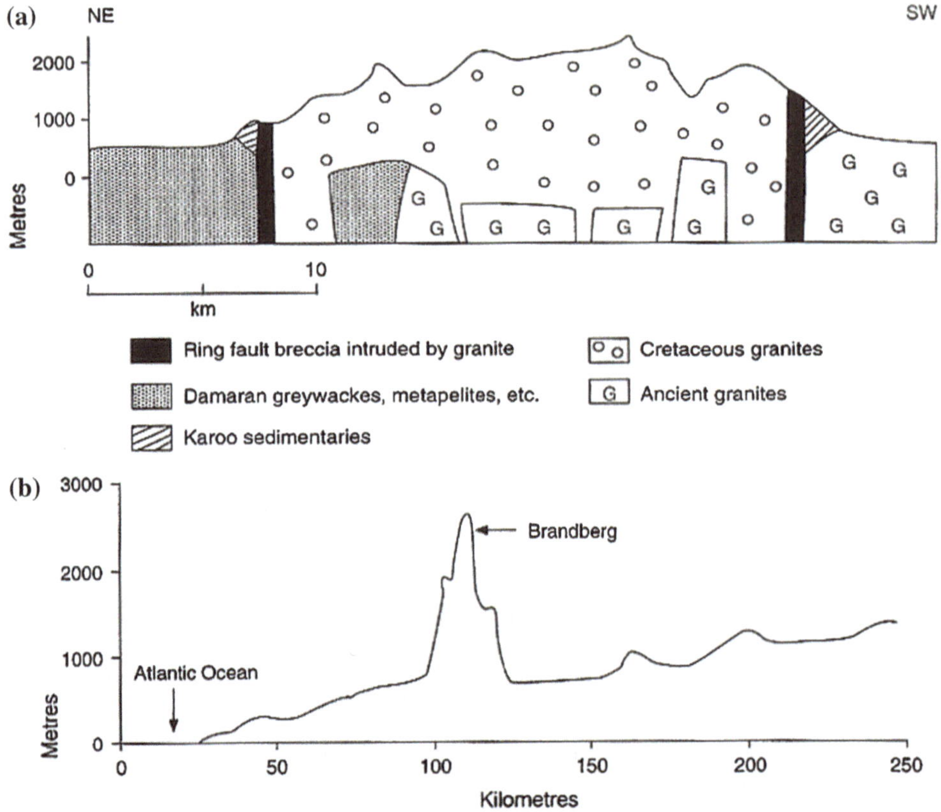

Fig. 10.2 Brandberg. **a** Geological cross-section modified from the Geological Map of Namibia, 1:250,000 series, sheet 2114, Omaruru (from Goudie and Eckardt 1999, Fig. 8), **b** Relief transect across Brandberg

10.2 Messum

The nature of the Messum ring complex (Fig. 10.4), named after sea captain and explorer William Messum, was unravelled by two great German geologists, Hermann Korn and Henno Martin in the 1930s. It lies to the southwest of Brandberg, covers approximately 400 km^2, has a diameter of approximately 18 km, and is bounded by ring-faults (Milner and Ewart 1989; Ewart et al. 2002). The geomorphology of the complex is striking and heavily influenced by the complex geology (Fig. 10.5), with the whole concentric structure recalling "petrified waves in a mud puddle into whose middle a stone has been dropped" (Korn and Martin 1954, p. 87). It consists of concentric hills and flats joined by radial gaps, with the outer rings being dominated by the black hues

Fig. 10.3 Google Earth image of Brandberg. Scale bar is 10 km (© 2012 CNES/Spot Image, Google)

Fig. 10.4 Google Earth image of Messum. Scale bar is 5 km (© 2012 GeoEye, Google, Digital Globe)

of gabbros and the more central parts by the grey tones of granites and syenites. Dating back to 135–132 million years, it is of similar age to features such as Erongo and the Etendeka volcanic sequences (Bauer et al. 2003). It was originally the centre of a volcanic caldera, formed when a volcano collapses down into itself (Ewart et al. 2002).

10.3 Spitzkoppe

Gross and Klein Spitzkoppe and the Pondok Mountains in west-central Namibia are dramatically prominent inselberg groups that are visible for huge distances (Fig. 10.6).

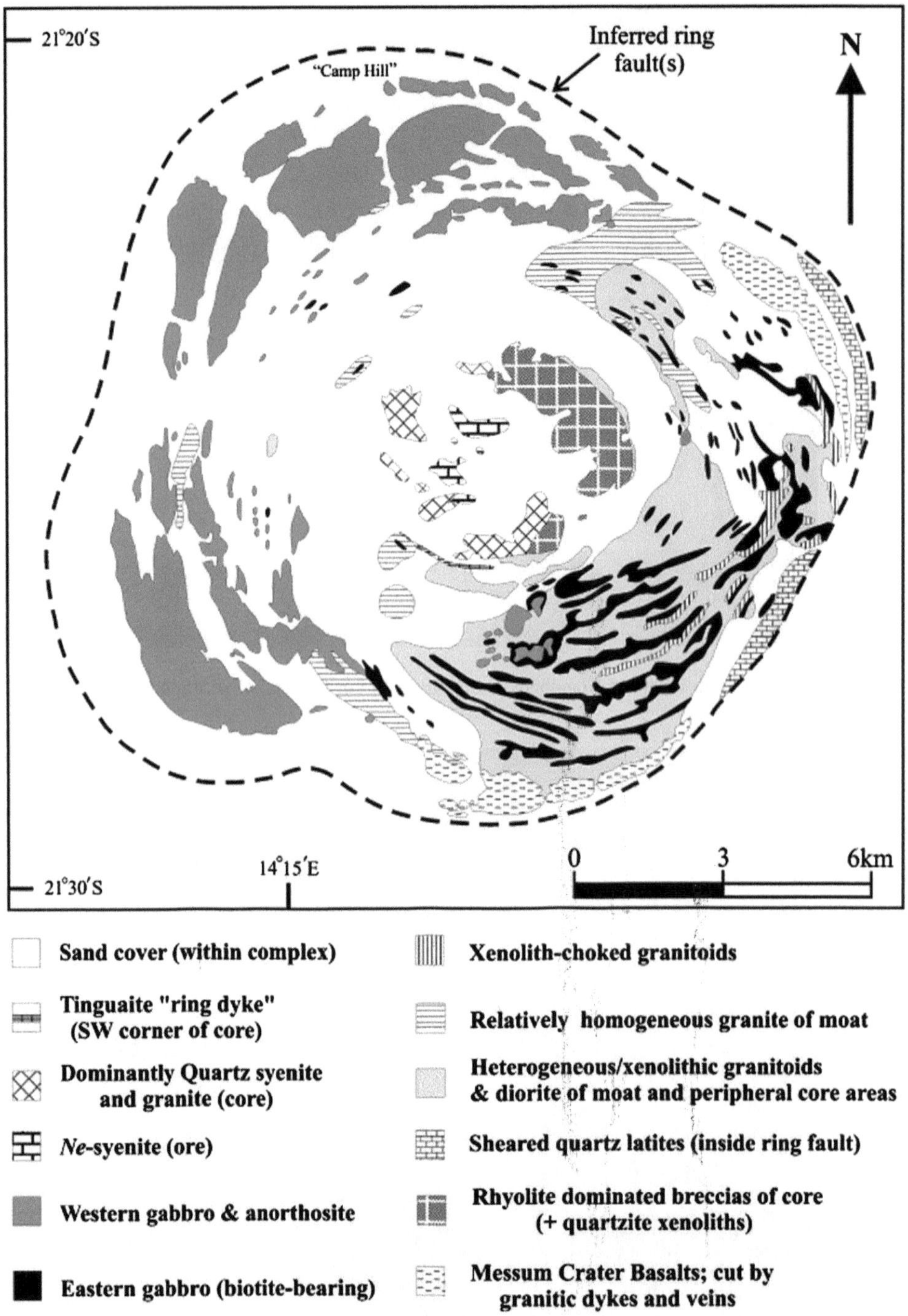

Fig. 10.5 The Geology of Messum Crater (from Ewart et al. 2002, Fig. 3)

Inselbergs, one of the iconic landforms of Africa, are isolated hills that stand above a flat or gently rolling topography. The topographic boundary between the hill and the plain is fairly abrupt. They are residual landforms made of strong and resistant rock, often granite, which have been developed as a result of the relative wearing down or back of the surrounding terrain. Spitzkoppe is one of the tallest, if not *the* tallest inselbergs on Earth (Migoń 2010) and is often referred to as the "Matterhorn of Africa". The gently sloping surfaces around the hills are called pediments. These are cut into rock and typically slope at less than 2°.

There have over the years been a range of hypotheses evoked to explain inselberg formation. One widely accepted view is that inselbergs are products of a two-stage development involving differential deep weathering in the first phase and stripping of the weathered mantle in the second one, which leaves an unweathered rock mass (the inselberg) at the surface. The unevenness in the depth of weathering results from differences in the degree of jointing in the bedrock. Another hypothesis, championed in southern Africa by Lester King (King 1949), invokes scarp retreat across unweathered, but possibly differentially jointed bedrock,

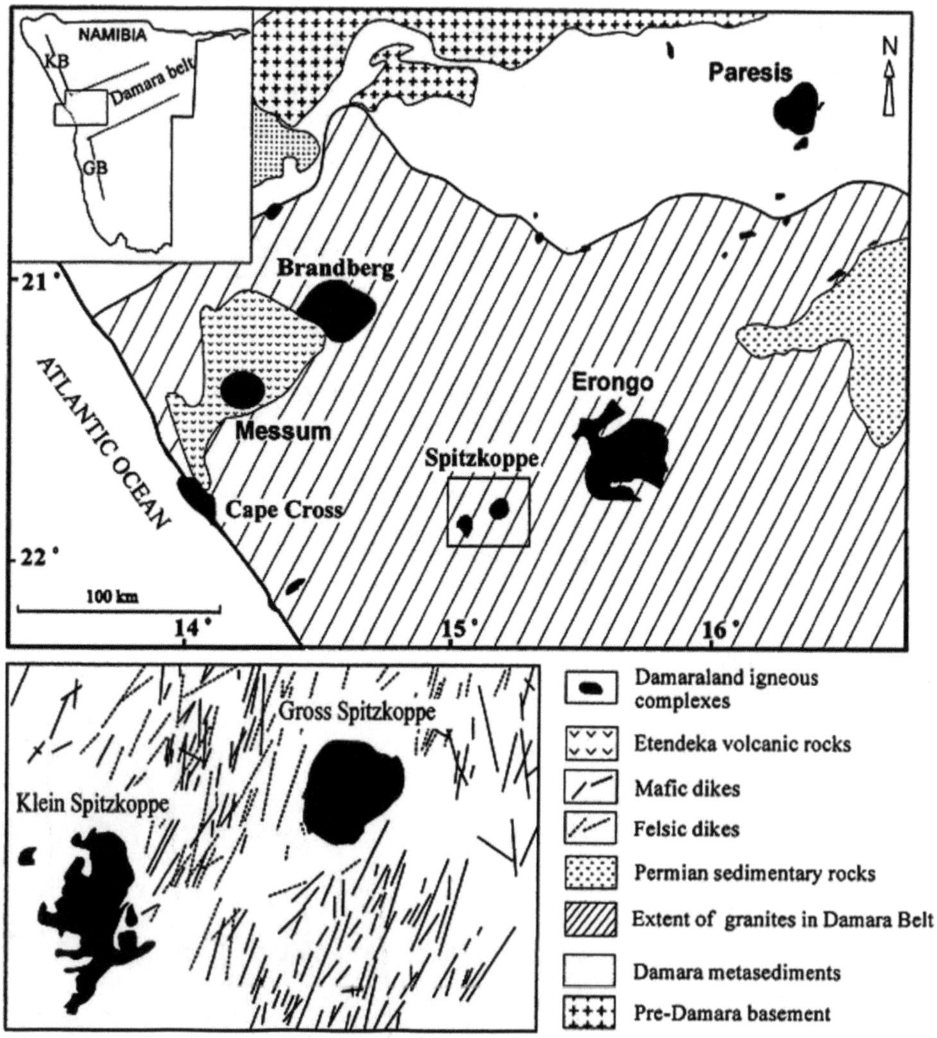

Fig. 10.6 Spitzkoppe and neighbours (after Frindt et al. 2004, Fig. 1)

leading to the creation of outliers at a varying distance from major escarpments. A third hypothesis, which seems widely applicable in Namibia, is that the inselbergs are the result of long continued differential erosion, with massively jointed rocks made of resistant minerals standing proud as the land around them is lowered (e.g. Selby 1982). In this respect, Spitzkoppe is composed of potassium-rich granite which is relatively resistant to chemical attack, mechanically strong, poorly jointed and with widely spaced fractures.

Recently, cosmogenic nuclides have been used to assess the rate at which the Spitzkoppe granites are eroding (Matmon et al. 2013). These suggest that the inselbergs are being lowered very slowly—at about 1–2 mm per thousand years, but that cliff retreat is taking place at a faster rate—c 8 mm per thousand years.

Gross and Klein Spitzkoppe rise abruptly and spectacularly from the partially calcreted planation surface developed across basement rocks, of which the Salem Granite and Damara mica schists and gneisses are the most important. Gross Spitzkoppe (Fig. 10.7) reaches an altitude of 1,728 m, Klein Spitzkoppe of 1,584 m and the Pondoks of 1,628 m, while the surrounding plains have an altitude of 1,000–1,100 m.

On the Spitzkoppe inselbergs a wide range of smaller landforms are present, largely produced by weathering processes. For example, weathering pits are extensively developed and unusually large (Goudie and Migoń 1997) as are rock shelters (as at Bushman's Paradise), and natural arches (e.g. The Bridge) (Fig. 10.8). The southwestern face of Gross Spitzkoppe also shows the effects of large-scale rock slides, with a chaos of enormous boulders, some as much as 30 m long, lying beneath an area from which a great slab, which was at least 50 m thick, has evidently been dislodged (Migoń 2006, p. 175). The Spitzkoppe granites are similar to those of Erongo and details of their composition are given by Mathias (1962), Frindt et al. (2004) and Haapala et al. (2007). They are of early Cretaceous age. The granites were emplaced at depths of several km below the ground surface that existed at the time, indicating substantial amounts of erosion since then. As indicated by the piles of boulders and rock debris surrounding the plains, massive rock falls have taken place.

Fig. 10.7 Gross Spitzkoppe

Fig. 10.8 Natural arch developed in granite at Spitzkoppe

References

Bauer K, Trumbull RB, Vietor T (2003) Geophysical images and a crustal model of intrusive structures beneath the Messum ring complex, Namibia. Earth Plan Sci Letters 216:65–80

Dauteuil O, Deschamps F, Bourgeois O, Mocquet A, Guillocheau F (2013) Post-breakup evolution and palaeotopography of the North Namibian Margin during the Meso-Cenozoic. Tectonophysics 589:103–115

Ewart A, Milner SC, Duncan AR, Bailey M (2002) The Cretaceous Messum igneous complex, S.W.Etendeka Namibia: reinterpretation in terms of a downsag-cauldron subsidence model. J Volcanol Geotherm Res 114:251–273

Frindt S, Trumbull RB, Romer RL (2004) Petrogenesis of the Gross Spitzkoppe topaz granite, central Namibia: a geochemical and Nd-Sr-Pb isotope study. Chem Geol 206:43–71

Goudie AS, Eckardt F (1999) The evolution of the morphological framework of the Central Namib Desert, Namibia, since the early Cretaceous. Geogr Ann 81A:443–458

Goudie AS, Migoń P (1997) Weathering pits in the Spitzkoppe area, Central Namib Desert. Z Geomorph 41:417–444

Haapala I, Frindt S, Kandara J (2007) Cretaceous Gross Spitzkoppe and Klein Spitzkoppe stocks in Namibia: Topaz-bearing A-type

granites related to continental rifting and mantle plume. Lithos 97:174–192
King LC (1949) A theory of bornhardts. Geogr J 113:83–87
Kirk-Spriggs AH, Marais E (2000) Dâures—biodiversity of the Brandberg Massif, Namibia. Cimbebasia Memoir 9, National Museum of Namibia, Windhoek
Korn H, Martin H (1954) The Messum igneous complex in south west Africa. Trans Geol Soc S Afr 57:83–122
Mathias M (1962) A disharmonious granite: the Spitzkop granite, south west Africa. Trans Geol Soc S Afr 65:281–292
Matmon A, Mushkin A, Enzel Y, Grodek T, ASTER Team (2013) Erosion of a granite inselberg, Gross Spitzkoppe, Namib Desert. Geomorphology 201:52–59
Migoń P (2006) Granite landscapes of the world. Oxford University Press, Oxford
Migoń P (2010) Spitzkoppe: the world of granite landforms. In: Migoń P (ed) Geomorphological landscapes of the world. Springer, Dordrecht, pp 155–162
Miller RMG (2000) Geology of the Brandberg Massif, Namibia and its environs. Cimbebasia Memoir 9:17–38
Miller RMG (2008) The geology of Namibia, vol 3. Geological Survey of Namibia, Windhoek
Milner SC, Ewart A (1989) The geology of the Goboboseb mountain volcanics and their relationship to the Messum Complex, Namibia. Comm Geol Surv Namibia 5:31–40
Raab MJ, Brown RW, Gallagher K, Weber K, Gleadow AJW (2005) Denudational and thermal history of the early Cretaceous Brandberg and Okenyenya igneous complexes on Namibia's Atlantic passive margin. Tectonics 24: TC 3006
Selby MJ (1982) Rock mass strength and the form of some inselbergs in the central Namib Desert. Earth Surf Proc Landf 7:489–497

11 Erongo

Abstract

Erongo, the largest of the Cretaceous plutons, is a caldera structure consisting of lavas and pyroclastic products. It was formed by cauldron subsidence, following which the granitic rocks were passively emplaced in the space provided by the subsidence. Where these granites outcrop, such as at Ameib, they produce a very striking landscape, with large domes, arches, precariously balanced boulders (as at the Bulls Parties), and remarkable weathering pits, while deep recesses produced by weathering have created numerous shallow caves or rock shelters.

The striking bulk of Erongo (Fig. 11.1), which lies some 25 km west of the small town of Omaruru, is the largest of the Cretaceous plutons, having a mean diameter of some 35–40 km. It was described, and its relative youth determined by the great German structural geologist Hans Cloos over 100 years ago (Korn and Martin 1953). Formed between 137 and 124 million years ago, with the greatest spasm of activity at c 132–130 million years ago (Wigand et al. 2003), it is contemporaneous with the Etendeka Lavas (see Chap. 9). It is irregular in outline, and is the only one of the plutons in which the roof strata are still substantially intact. It forms a large mass (Fig. 11.2) that rises to some 1,000 m above the plain and attains a maximum altitude of 2,305 m (Hüser 1977; Blümel et al. 1979). Karoo sediments outcrop as a band along the edge of the Erongo mountains and dip between 5° and 20° toward their centre (Hegenberger 1988). The interpretation of Erongo is that it is a central caldera structure consisting of lavas and pyroclastic products (Fig. 11.3). It was formed by cauldron subsidence, following which the granitic rocks were passively emplaced in the space provided by the subsidence (Pirajno 1990). Encircling the main complex is a ring dyke of olivine dolerite, up to 200 m thick.

The Cretaceous granites that outcrop at Ameib, which is at the southern edge of the Erongo mountains, produce a very striking landscape, with large domes, arches, precariously balanced boulders (as at the Bulls Parties) (Fig. 11.4), and remarkable weathering pits. Deep recesses produced by weathering have created numerous shallow caves or rock shelters in which rock paintings and other evidence of human occupation occur. The most famous of these is Phillip's Cave. The dome into which Phillip's Cave is cut is notable for some patterns that have developed on the surface of the granite (Fig. 11.5). Polygonal fracture patterns consist mainly of pentagonal or hexagonal cracks meeting at ~120° where they make tri-radial junctions on predominantly curved rock surface pavements (Young et al. 2009). Polygonal crack diameters may vary from ~5 cm to over 50 cm and in some instances micro-polygonal cracking has been observed within larger tessellation plates (Robinson and Williams 1989).

Although understanding the origin of tessellation has caused some debate (Young et al. 2009), and may differ between sites, it is a particularly common phenomenon in case-hardened rocks associated with regions of seasonal precipitation patterns (Robinson and Williams 1989, 1992). Some of the likely causative mechanisms suggested include shrinkage of silica gel due to changing rock thermal and/or moisture conditions (Robinson and Williams 1989, 1992), simply through thermal changes which set up differential surface stress (Branagan 1983; Croll 2009), expansion to accommodate precipitating solutions of iron, etc., and thermal stresses or geochemical changes caused by emplacement of dolerite dykes (Velázquez et al. 2008).

The weathering pits at the Bulls Parties are clearly visible on satellite images (Fig. 11.6). One of the pits is notable for being deep (c 4 m) yet only 6–7 m across. It is also remarkable for the fact that its floor is flat and devoid of any sediments, which brings about the question of what happened to the weathered material liberated during its

A. Goudie and H. Viles, *Landscapes and Landforms of Namibia*, World Geomorphological Landscapes,
DOI 10.1007/978-94-017-8020-9_11,

Fig. 11.1 Erongo

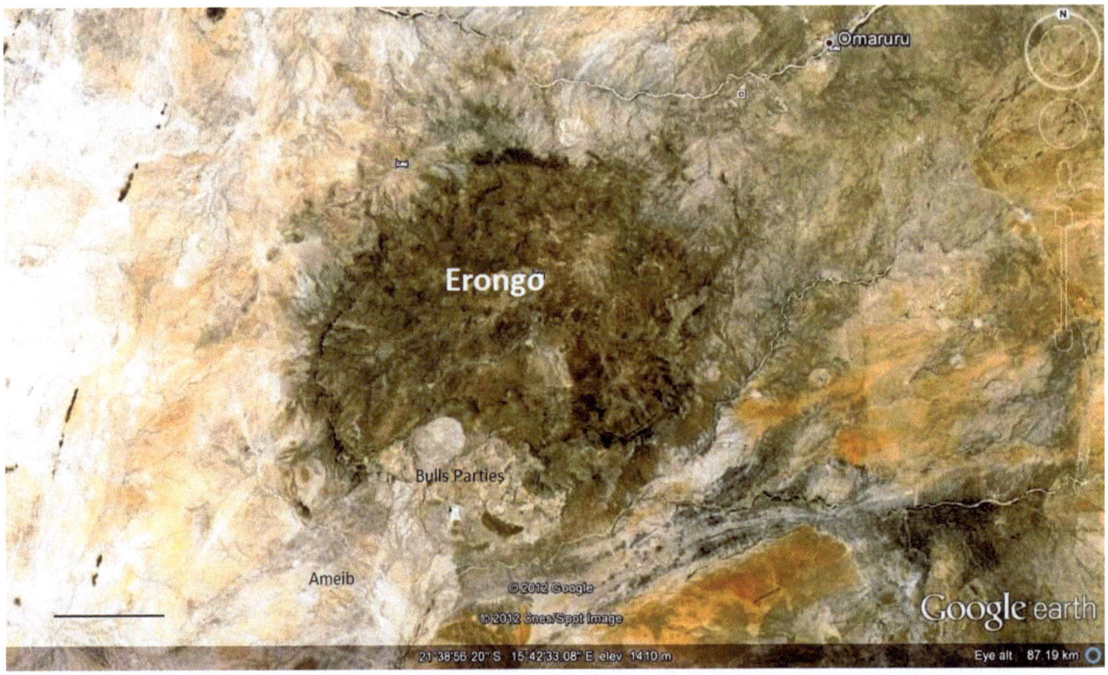

Fig. 11.2 Google Earth Image of Erongo. Scale bar 10 km (© 2012 Google, CNES/Spot Image)

formation (Goudie and Migon 1997). Three possible ways have been suggested (Smith 1941) to explain sediment-free pits (Fig. 11.7), namely solutional transport, washing out during excessive rainfalls, and deflation. Flotation is another possible mechanism, but no direct observations have been made. However, the occurrence of deep closed pits devoid of any sediment remains puzzling and no satisfactory explanation of their emptiness has been offered (Netoff et al. 1995). On the other hand some pits contain large numbers of flat stones, some of which have possibly been moved from surrounding slopes by scavenging baboons.

Such small closed depressions, also called 'gnammas', 'Opferkessel' or 'pias', are common on horizontal and gently inclined rock surfaces in many parts of the world's drylands. They have been described from a range of silicate rock types, most frequently granites (Twidale and Vidal Romani 2005) and sandstones (Young et al. 2009). They may be similar in their broad morphology to solutional pits

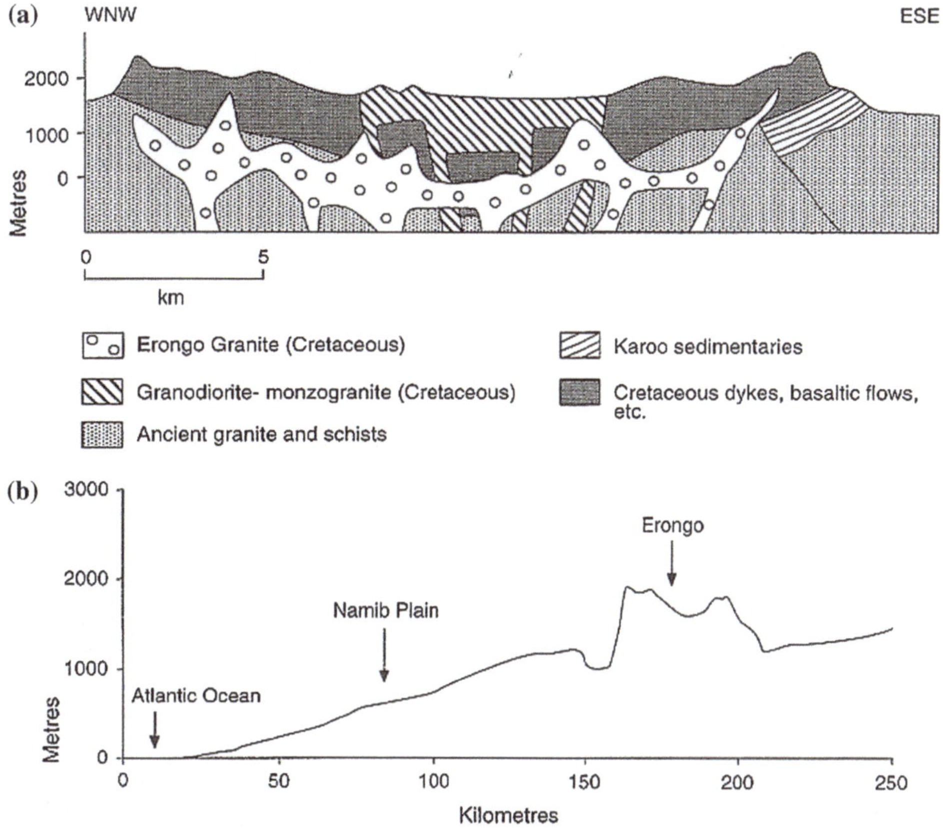

Fig. 11.3 Erongo. **a** Geological cross-section (modified from the Geological Map of Namibia, 1:250,000 series, sheet 2114, Omaruru). **b** Topographic situation (from Goudie and Eckardt 1999, Fig. 4)

Fig. 11.4 Granite boulders on pedestals at the Bulls Parties, Ameib

developed in carbonate rocks, to which the term 'kamenitza' is often applied. Pits may have a variety of forms, including pans, bowls, cylinders and armchairs.

There is a great deal of uncertainty concerning the processes involved in their development. Chemical processes of solution are usually invoked (Domínguez-Villar et al. 2008),

Fig. 11.5 Polygonal cracking in granite, Phillip's Cave

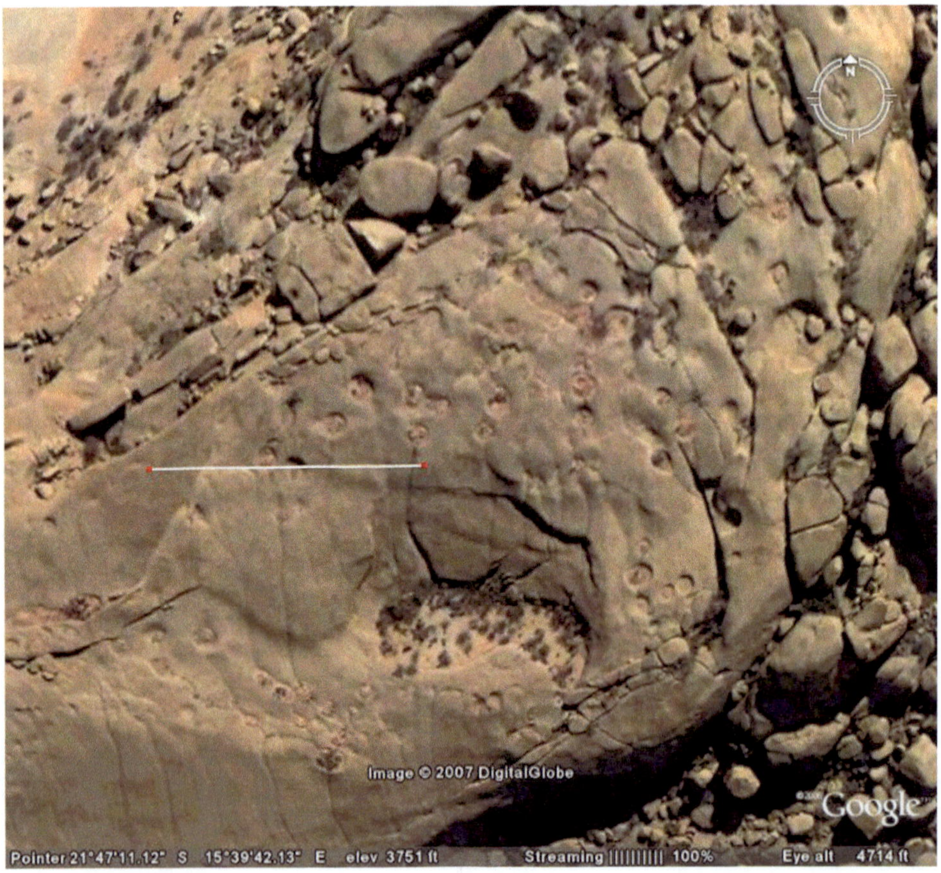

Fig. 11.6 Google Earth Image of the Bulls Parties, Ameib, showing weathering pits on dome surface. Scale bar 100 m (© 2007 Digital Globe)

but other processes may include hydration, the mechanical action of frost and salt and biochemical weathering. Complex biofilms, often black in colour, may accumulate at the base of rock basins; these are known to dissolve the cement between sandstone grains and also act as biological sealants to water infiltration (Chan et al. 2005). Positive feedback mechanisms

Fig. 11.7 Weathering pit, Erongo. Notice the lack of sediment. Pole for scale

related to the ever-growing amount of water available as a pit enlarges may account for a localised high intensity of weathering (Schipull 1978). Goudie and Migoń (1997) argued that weathering pits are of larger-scale significance, hypothesising that they played a major role in the development of the pediments that surround the inselbergs, as their lateral and vertical growth contributes to surface lowering.

References

Blümel WD, Emmermann R, Hüser K (1979) Der Erongo. Geowissenschaftliche Beschreibung und Deutung eines südwestafrikanischen Vulkankomplexes. Sci Res Series SWA Series, SW Afr Sci Soc 16

Branagan DF (1983) The Sydney basin and its vanished sequence. J Geol Soc Austr 30:75–84

Chan MA, Beitler Bowen B, Parry BWT, Ormö J, Komatsu G (2005). Red rock and red planet diagenesis: comparisons of Earth and Mars concretions. GSA Today 15:4–10

Croll JGA (2009) Possible role of thermal ratchetting in alligator cracking of asphalt pavements. Int J Pavement Eng 10:447–453

Domínguez-Villar D, Arteaga C, Garcia-Gimenez R, Smith E, Pedraza J (2008) Diurnal and seasonal water variations of temperature, pH, redox potential and conductivity in gnammas (weathering pits): implications for chemical weathering. Catena 72:37–48

Goudie AS, Eckardt F (1999) The evolution of the morphological framework of the Central Namib Desert, Namibia, since the early Cretaceous. Geogr Ann 81A:443–458

Goudie AS, Migoń P (1997) Weathering pits in the Spitzkoppe area, central Namib Desert. Z Geomorph 41:417–444

Hegenberger W (1988) Karoo sediments of the Erongo Mountains, their environmental setting and correlation. Comm Geol Surv South West Africa/Namibia 4:51–57

Hüser K (1977) Namibrand und Erongo. Karlsruher Geogr Hefte 9

Korn H, Martin H (1953) Der intrusionmechanismus der Grossen Karroo-Plutone in Südwestafrika. Geol Rundsch 41:41–58

Netoff DI, Cooper BJ, Shroba RR (1995) Giant sandstone weathering pits near Cookie Jar Butte, Southeastern Utah. Trans and Proc Series NPS/NRNAU/NRJP 95(11):25–53

Pirajno F (1990) Geology, geochemistry and mineralisation of the Erongo volcanic complex, Namibia. S Afr J Geol 93:485–504

Robinson DA, Williams RBG (1989) Polygonal cracking of sandstone at Fontainbleau, France. Z Geomorph 33:59–72

Robinson DA, Williams RBG (1992) Sandstone weathering in the high Atlas, Morocco. Z Geomorph 36:413–429

Schipull KV (1978) Waterpockets (Opferkessel) in Sandsteinen des zentralen Colorado-Plateaus. Z Geomorph 22:426–438

Smith LL (1941) Weather pits in granite of the southern Piedmont. J Geomorph 4:117–127

Twidale CR, Vidal Romani JR (2005) Landforms and geology of granite terrains. Balkema, Leiden

Velázquez VF,Giannini PCF, Riccomini C, Sallun AEM, Hachir J, Gomes C de B (2008) Columnar joints in the Patiño formation sandstones, eastern Paraguay: a dynamic interaction between dyke intrusion, quartz dissolution and cooling-induced fractures. Episodes 31:302–308

Wigand M, Schmitt AK, Trumbull RB, Villa IM, Emmermann R (2003) Short-lived magmatic activity in an anorogenic subvolcanic complex: $^{40}Ar/^{39}Ar$ and ion microprobe U-Pb dating of the Erongo, Damaraland, Namibia. J Volcanol Geothermal Res 2715:1–21

Young R, Wray R, Young A (2009) Sandstone landforms. Cambridge University Press, Cambridge

Weathering on the Namib Plains: Marble and Granite

12

Abstract

In the central Namib there are numerous outcrops of Pre-Cambrian dolomitic marbles. Mechanical disintegration appears to be the most widespread weathering process operating on this marble, producing extensive spreads of coarse marble sand akin to granitic grus. On the other hand, some marble outcrops, including those at the very foggy location of Swartbank, have solutional rillenkarren, particularly on rock faces oriented towards the Atlantic Ocean and the direction of incoming fogs. Some marble outcrops display the impact of lichen weathering. East-facing marble surfaces within the central Namib also exhibit a wide range of aeolian fluting, grooves and helical scores. The many granite outcrops of the area display a fine array of weathering forms, including tafoni, accumulations of grains (*grus*), A-tents, pits, exfoliation features, and rinds.

12.1 Introduction

Rising up above the Namib Plains between the Swakop and Kuiseb rivers, are a number of inselbergs composed of marbles (e.g. the Hamilton Range) and granites (e.g. Mirabib and Vogelfederberg). The hills and the plains around them are an excellent environment in which to study weathering forms and processes. Weathering forms abound and they impressed the Victorian explorer and painter Thomas Baines (1864, p. 23) as he passed up the Swakop valley on his way from Walvis Bay to the interior:

> Passing through a small break in the hills we entered a small desolate valley enclosed by barren pyramids, cones and precipices of fantastic shapes and dry arid colours, the yellowish grey of the generality of the stone being only relieved by a darker tint banded by light pink streaks of quartz crossed by lodes of black ironstone, or speckled by a black substance splitting easily into thin glittering laminae. The whole surface of the rock seemed undergoing rapid disintegration and in places it was almost dangerous to step on what seemed a solid rock lest it should crumble under foot. In this manner, caves, holes in the rocks and peaks or blocks of every shape were found. One which I sketched had a singular resemblance to a gigantic head and face and others Mr Dixon told me were called Hansom Cab and other names expressive of their shape.

12.2 Marble

In the central Namib there are numerous outcrops of Pre-Cambrian dolomitic marbles. They weather and break down in a range of different ways. While it is possible that various mechanisms could contribute to this degradation, Rayleigh (1934) suggested that marbles that had been baked at temperatures of 100 °C or even less had their rigidity diminished. He suggested that this was due to the uneven way in which calcite crystals expanded on heating. Rayleigh was, therefore, advocating heating and cooling (insolation weathering or thermoclasty) as a cause of marble breakdown. More recently, Royer-Carfagni (1999) has used scanning electron microscopy to identify changes in marble structure on heating, and has suggested why it is that certain types of marble may be prone to the effects of quite modest heating treatments (p. 119): 'Calcite is known to expand on heating much more in the direction of its optical axis than perpendicular to it. The grains' shapes change with temperature and a grain which fits snugly into the mosaic at a given temperature is no longer able to do so when the temperature is varied; this is because the anisotropy directions of the individual grains are oriented randomly. The

A. Goudie and H. Viles, *Landscapes and Landforms of Namibia*, World Geomorphological Landscapes,
DOI 10.1007/978-94-017-8020-9_12,

Fig. 12.1 Marble rillen at Swartbankberg in the Central Namib

result is a springing apart of contiguous grains, giving rise to a non-zero residual stress state inside the material.'

Sweeting and Lancaster (1982) indicated that mechanical disintegration appears to be the most widespread weathering process operating on marble within the central Namib, producing extensive spreads of coarse marble sand like granitic grus. Near coastal salt pan environments, blocks cut from locally-sourced Karibib marble in a field experiment showed deterioration after only 3 years in situ as a result of salt weathering (Viles and Goudie 2007). However, Viles (2005, p. 205) emplaced locally-quarried Karibib marble blocks at four locations (Kleinberg, Vogelfederberg, Gobabeb and Ganab) which showed much less deterioration than those left out in the coastal salt pan. After three years, SEM observations showed that "the marble blocks were undergoing internal weakening through widening of grain boundaries and occasional cracking of calcite crystals". This was very similar to the effects produced by experimental heating and cooling of marble blocks in the laboratory (Goudie and Viles 2000). Blocks from Ganab (over 100 km inland from the coast) were the most affected and crumbled easily to granular debris when sub-sampled with a geological hammer, and this was the site where the highest diurnal rock surface temperature ranges were recorded with a data logger. Viles (2005, p. 205) concluded, "In the absence of any SEM evidence for dissolution, biological weathering or salt crystallization occurring on the marble blocks, thermal expansion and contraction producing fatigue and crack development is proposed as the most likely process occurring during the initial stages of breakdown".

On the other hand, some marble outcrops, including those at the very foggy location of Swartbank near the Kuiseb River downstream from Gobabeb, have solutional rillenkarren (Fig. 12.1), particularly on rock faces oriented towards the Atlantic Ocean and the direction of incoming fogs (Sweeting and Lancaster 1982, p. 200). At a larger scale, there are some marbles in the central Namib that are the sites of caves, but on account of the aridity this speleogenesis has been attributed to hydrothermal action from underground water sources (Irish et al. 2000).

Some marble outcrops display the impact of lichen weathering. At Swartbank, for example, Viles and Goudie (2000) collected samples that under the SEM showed borehole production by fungal hyphae at the base of lichens—illustrating the importance of biochemical weathering processes in areas which experience frequent fogs. Indeed, within the fog belt of the coastal Namib Desert a diverse lichen community is found (especially up to about 30 km from the coast) where surfaces are stable enough to support them. According to Schieferstein and Loris (1992) maximum lichen coverage is found about 5 km from the coast, whilst biomass peaks at c 1 km from the coast. Foliose and fruticose species are often found on west-facing surfaces in the central Namib, whilst crustose (and especially endolithic) species inhabit east-facing surfaces which are prone to drying winds coming from the east (Schieferstein and Loris 1992).

Lichens weather rocks through a range of biochemical and biophysical processes, often working synergistically. However, lichens can also provide a bioprotective role on rock surfaces, reducing the impact of other weathering processes and enhancing surface stability (Viles 2008). Lichens growing in desert and other hostile environments are now known to be associated with a wide range of microbial species as well as arthropods and other small animals (Lalley and Viles 2005), and such micro-ecosystems may contribute to weathering in complex ways.

Finally, east-facing marble surfaces within the central Namib also exhibit a wide range of aeolian fluting, grooves and helical scores (Bourke and Viles 2007). Strong, dry berg winds blow intermittently from the east, bringing sand which can then abrade east facing rock surfaces. Abrasion results in a great diversity of small scale relief features (Fig. 12.2) depending on the structure and texture of the rocks involved. There has been much debate about how and why regular abrasion features such as aeolian flutes form as reviewed by Bourke and Viles (2007).

Thus, across the central Namib there is evidence for multiple processes of weathering and rock breakdown affecting marble outcrops, working together to produce a great diversity of small scale features. A general gradient can be observed from salt weathering dominating in pans near the coast, biochemical and dissolutional effects and wind fluting around Gobabeb, Swartbank and Vogelfederberg and thermal insolation weathering further inland.

Fig. 12.2 Aeolian abrasion features on marble exposures in the Central Namib

12.3 Granite

The many granite outcrops of the central Namib display a fine array of weathering forms, including tafoni (Fig. 12.3), accumulations of grains (*grus*), A-tents, pits, exfoliation features, split boulders and rinds (Selby 1977b; Ollier 1978; Migoń and Goudie 2000). Many of these features are beautifully displayed in the grounds of the Desert Training and Research Station at Gobabeb and its immediate environs (Goudie 1972) (Fig. 12.4).

Granites are found widely across the region, including the Damara granite dating from 470 to 650 Myr ago, and form the main inselbergs and peaks within the area e.g. Mirabib (Fig. 12.5). The mechanisms by which the granitic inselbergs in Namibia themselves came into existence have been discussed by Selby (1977a, b; 1982) and Ollier (1978) (see Chap. 11 for more details).

These granites are extremely hard and resilient to weathering, as evidenced by the lack of deterioration of Damara granite blocks after 2 years in situ in a coastal salt pan and complete absence of evidence of any weathering after 3 years exposed at inland sites in comparison with Karibib marble blocks (Viles 2005; Viles and Goudie 2007). Furthermore, most granite outcrops across the central Namib exhibit surface crusting with an often hard, dark brown (iron-rich) layer a few mm to cm thick. There appears to be some climatic control of this crust development, with particularly clear examples further east where higher rainfall is experienced. At Gobabeb one of the intriguing weathering features is the presence of iron-enriched raised rims left after weathering of small granite outcrops (Migoń 2006, p. 284).

Vogelfederberg inselberg (Fig. 12.6), located some 60 km east of Walvis Bay, is a relatively low granite dome, about 50 m high. It is notable, however, for having a partially sand-filled, shallow depression present around a portion of its perimeter. Such features are known as scarp-foot depressions, or *Bergfussniederungen* in German. Migoń (2006, p. 115) suggests that their origin 'is related to subsurface weathering, which is enhanced at the slope/plain junction because of increased availability of water, derived from surface runoff'. In addition he suggests that 'Episodic washing of weathering material excavated an initial trough, which may deepen through both subsequent weathering and episodic fluvial erosion'.

Tafoni and other cavernous weathering features are also found on many granite outcrops within the central Namib Desert, including notable occurrences on the granite boulders which mantle the surfaces surrounding Gobabeb and on the sides of Mirabib. Tafoni (singular *tafone*) are cavernous

Fig. 12.3 Tafoni underming the granite at Mirabib inselberg

Fig. 12.4 Split boulders at Gobabeb

weathering forms (Fig. 12.7) typically several cubic metres in volume with arch-shaped entrances, concave inner walls, overhanging margins (visors) and fairly smooth gently sloping, debris-covered floors. First described from Corsica, they occur in many parts of the world (Goudie and Viles 1997), including polar regions, and have been described in many deserts.

These enigmatic features occur where some kind of positive feedback sets in as a small initial hollow becomes enlarged, thus altering the microclimatic conditions within it and further encouraging weathering. Around the hollow, the development of surface crusts probably also inhibits such weathering further focusing weathering activity differentially within the expanding hollow. The cavernous hollows of

Fig. 12.5 Google Earth image of Mirabib inselberg. Scale bar 0.1 km (© 2012 GeoEye, Google)

Fig. 12.6 The Vogelfederberg inselberg

tafoni are believed to result largely from flaking and granular disintegration caused by a range of possible weathering processes that include hydration, salt crystallization, lichen growth, and chemical attack by saline solutions. Some workers have found clear evidence of salt weathering being involved, while others have not. The role of case hardening in their formation is also the subject of debate, but can help to explain the formation of the visor. For a cavity to grow there needs to be a mechanism to remove flakes and spalls. Wind may play a part, as do organisms such as pack rats. Although some early workers thought that the actual excavation of a cavity might be achieved by wind abrasion, many tafoni occur in environments where sand blasting does not occur or they may have an aspect (i.e. the leeward side of a boulder) or a height up a cliff face that precludes such a mechanism. Further research is needed to both understand the general positive feedback loops which occur, and the contribution of individual processes in order to explain in more detail why tafoni develop, how quickly they form, and whether they are still actively developing in the coastal Namib Desert.

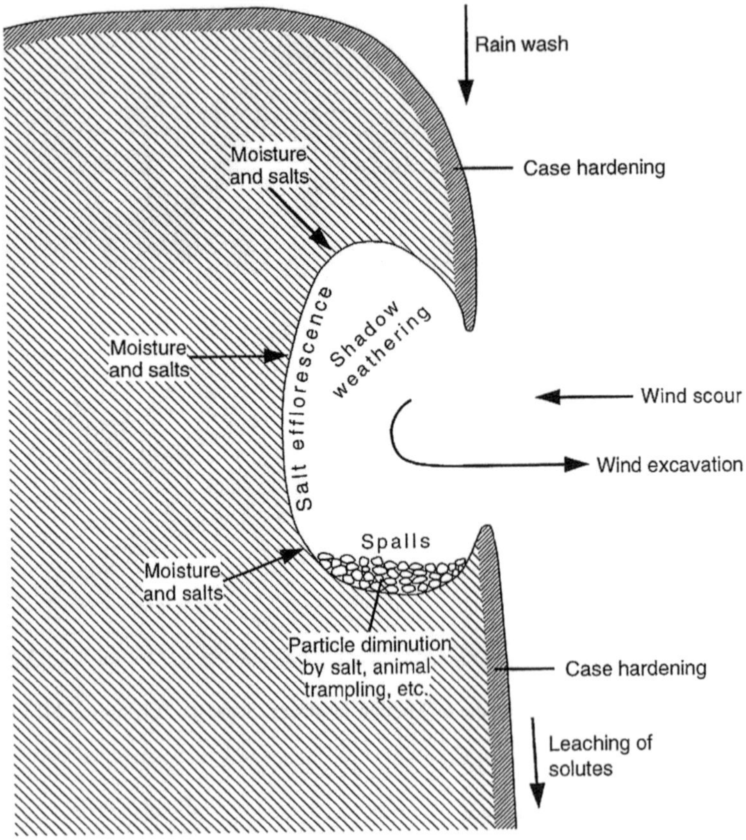

Fig. 12.7 A model of tafoni development (from Goudie and Viles 1997, Fig. 6.9)

References

Baines T (1864) Explorations in South-West Africa. Longman Green, London

Bourke MC, Viles HA (2007) A photographic atlas of rock breakdown features in geomorphic environments. Planetary Science Institute, Tucson

Goudie A (1972) Climate, weathering, crust formation, dunes and fluvial features, of the Central Namib Desert, near Gobabeb, South West Africa. Madoqua Ser II 1:15–31

Goudie AS, Viles HA (1997) Salt weathering hazard. Wiley, Chichester

Goudie AS, Viles HA (2000) The thermal degradation of marble. Acta Universitas Carolinae 2000 Geographica, XXXV, Supplementum, pp 7–16

Irish J, Martini J, Marais E (2000) Cave investigations in Namibia VI. The desert caves in and around the Namib-Naukluft Park. Cimbebasia 16:177–193

Lalley JS, Viles HA (2005) Terricolous lichens in the northern Namib Desert of Namibia: distribution and community composition. Lichenologist 37:77–92

Migoń P (2006) Granite landscapes of the world. Oxford University Press, Oxford

Migoń P, Goudie A (2000) Granite landforms of the Central Namib. Acta Universitatis Carolinae 2000 Geographica, XXXV, Supplementum, pp 17–38

Ollier CD (1978) Inselbergs of the Namib Desert. Processes and history. Z Geomorph Suppl 31:161–176

Rayleigh L (1934) The bending of marble. Proc Roy Soc 144:266–279

Royer-Carfagni GF (1999) On the thermal degradation of granite. Int J Rock Mech Mining Sci 36:119–126

Schieferstein B, Loris K (1992) Ecological investigations on lichen fields in the Central Namib 1: distribution patterns and habitat conditions. Vegetatio 98:113–128

Selby MJ (1977a) Bornhardts of the Namib Desert. Z Geomorph NF 21:1–13

Selby MJ (1977b) On the origin of sheeting and laminae in granitic rocks: evidence from Antarctica, the Namib Desert and the Central Sahara. Madoqua 10:171–179

Selby MJ (1982) Rock mass strength and the form of some inselbergs in the Central Namib Desert. Earth Surf Proc Landf 7:489–497

Sweeting MM, Lancaster N (1982) Solutional and wind erosion forms on limestone in the Central Namib Desert. Z Geomorph NF 26:197–207

Viles HA (2005) Microclimate and weathering in the Central Namib Desert, Namibia. Geomorphology 67:189–209

Viles HA (2008) Understanding dryland landscape dynamics: do biological crusts hold the key? Geogr Compass 2(3):899–919

Viles HA, Goudie AS (2000) Weathering, geomorphology and climatic variability in the Central Namib Desert. In: McLaren SJ, Kniveton DR (eds) Linking climate change to land surface change. Kluwer, Dordrecht, pp 65–82

Viles HA, Goudie AS (2007) Rapid salt weathering in the coastal Namib desert: implications for landscape development. Geomorphology 85:49–62

Salt Weathering in the Namib: Soutrivier and the Coastal Salt Pans

13

Abstract

In the coastal zone of the Namib salt weathering is a potent agent of rock disintegration. It has also caused the corrosion and failure of various engineering structures. Spectacular examples of the impacts of salt weathering can be seen along the coastal road from Swakopmund to Henties Bay, and where the road from Vogelfederberg to Gobabeb crosses the Soutrivier. At Soutrivier there are salt crusts and efflorescences in and around the damp river bed, and the granites show a wide range of weathering phenomena, including splitting, flaking, alveoles and tafoni. Not only has salt weathering given rise to extensive weathering phenomena but it may help to explain the development of the central Namib plains by a process of 'haloplanation', generating the great topographic monotony of some of the Namib coastal plain areas. Furthermore, salt weathering may be an important source of dust which blows off the Namib coast in great plumes. Recently, various experiments based on field exposure of rock blocks over one or more years have demonstrated the rapidity of presumed salt weathering in the Namib. These studies using rock blocks have confirmed that highly aggressive ground conditions occur in the fog belt and within and around coastal salt pans, producing weathering 'hotspots'. Further inland conditions are less severe and salt weathering is limited to topographic hollows where suitable microclimatic conditions prevail and where groundwater seepage occurs.

13.1 Introduction

One consequence of the presence of substantial quantities of both calcium sulphate and sodium chloride in the coastal zone of the Namib is that salt weathering is a potent agent of rock disintegration (Viles and Goudie 2013). It has also caused the corrosion and failure of various engineering structures including the pipeline that supplies water from the Omaruru River aquifer to Swakopmund and Rössing (Bulley 1983) (Fig. 13.1). Signs of the potency of salt weathering can be seen in many places within the coastal Namib Desert, such as where the road from Vogelfederberg to Gobabeb crosses the aptly named, Soutrivier (Salt River) (Fig. 13.2). Here there are salt crusts and efflorescences in and around the damp river bed, and the granites show a wide range of weathering phenomena, including splitting, flaking, alveoles and tafoni (Fig. 13.3).

13.2 Salt Weathering Mechanisms and Implications

It has been known since antiquity that salt attacks rock and building materials (Goudie and Viles 1997), but major developments in the study of salt weathering only came from the end of the nineteenth century onwards. It has become clear that salt weathering comprises a range of mechanisms, some chemical and some mechanical. Experimental studies have demonstrated that, of the mechanical processes, crystal growth from solution in rock pores and cracks is the most effective. It results from a decrease in solubility as temperature falls, by evaporation of solutions, or by mixing of different salts in solution (the common ion effect).

However, disruptive stresses may also be exerted by anhydrous salts, dehydrated in high desert temperatures, which from time to time become hydrated. Sodium sulphate,

A. Goudie and H. Viles, *Landscapes and Landforms of Namibia*, World Geomorphological Landscapes,
DOI 10.1007/978-94-017-8020-9_13,

Fig. 13.1 Water pipeline ruined by salt attack between Rössing and Swakopmund

Fig. 13.2 Soutrivier

sodium carbonate and magnesium sulphate are examples of these sorts of salts. As a change of phase takes place to the hydrated form, water is absorbed. This increases the volumes of the salt and thus develops pressure against pore walls. In addition, many common desert salts have high coefficients of volumetric expansion that are higher than those of common rocks such as granite (Cooke and Smalley 1968). Their expansion may set up disruptive stresses in the rock.

In addition to these three main categories of mechanical effects, salt can also cause chemical weathering in susceptible rocks (Schiavon et al. 1995), and experimental studies using a range of salts and scanning electron microscopy showed

Fig. 13.3 Soutrivier, showing weathering forms

that salts could cause chemical etching of quartz grains after only short periods of exposure (Magee et al. 1988).

Not only has salt weathering given rise to extensive relatively small scale landforms such as weathering phenomena but it may help to explain the development of the central Namib plains by a process of 'haloplanation', generating the great topographic monotony of some of the Namib coastal plain areas (as for example between Uis and Henties Bay). Evidence for this haloplanation action comes from observations of rocks suffering from what has been termed 'shark fin weathering', in that a narow fin is left protruding above the ground surface as a remnant of a much larger sub-surface clast (Fig. 13.4). Over time, the fins will disappear and the entire land surface becomes planed off. Salt weathering may also explain the high rates of denudation indicated by some cosmogenic isotope studies of inselbergs. Cockburn et al. (1999) used measurements of cosmogenic ^{10}Be and ^{26}Al from granite inselbergs at three central Namib locations at 40–80 km from the coast—Blutkopje, Mirabib and Vogelfederberg to estimate long term erosion rates. A mean rate of summit lowering of 5.07 ± 1.1 m per million years over the last >10^5 years was recorded. In comparison with data from Australian inselbergs (where rates of around 0.7 m per million years have been calculated) this is quite rapid. They attributed this higher denudation rate in the central Namib to active salt weathering associated with ample fog precipitation at the sampled inselbergs.

One link between weathering and erosion of specific interest which links the landscape of the central Namib to the global climate system concerns the interplay of salt weathering and dust production. Great plumes of dust can often be seen blowing off the Namibian desert in MODIS and other satellite imagery (e.g. Viles and Goudie 2007, Fig. 9). Weathering in salt pans is one key source of such fine-grained debris, which plays an important role in soil development and biogeochemical cycles within the desert environment and can also be entrained in upper winds and carried across the Atlantic. The fact that there are several semi-permanent

Fig. 13.4 Two examples of 'shark fin weathering' from the moist salty zone at Tomato Pan, north of Swakopmund

groundwater seeps across the central Namib, which are likely to experience similarly high rates of salt weathering, means that such weathering hotspots are highly likely to be important contributors to the dust budget here.

13.3 Weathering Experiments: The Role of Moisture

Recently, various experiments based on field exposure over one or more years of rock blocks have demonstrated the efficacy of salt weathering in the Namib. Goudie et al. (1997) emplaced blocks of a Jurassic oolitic limestone in a near-coastal salt pan (Tomato Pan) within the fog belt 22 km north of Swakopmund and at a distance of c 4 km from the coast. The land surface there is often moist and is impregnated with halite (NaCl) and gypsum. Rock outcrops display flakes, splits, alveoli, shark fin weathering etc. The rock blocks were left in this environment for 2 years. Many displayed considerable disintegration, particularly those that had been emplaced on salty, stone pavement surfaces. Geochemical analyses of the weathered blocks suggested that the active salt was halite and the climatology of the area implies that numerous salt crystallization cycles occurred in response to frequent wetting of rocks by fogs and subsequent drying by the sun and the wind. Subsequent laboratory simulations under Namib conditions confirmed the effectiveness of sodium chloride crystallization in such a coastal, foggy, pavement environment (Goudie and Parker 1998).

Viles and Goudie (2007) emplaced a wider range of rock types, including two Namibian rocks, Karibib marble and Damara granite, at the same Tomato Pan site. After 2 years of exposure the pre-weighed cut blocks were reweighed, tested for strength (Dynamic Young's Modulus) using the Grindosonic apparatus, examined under the SEM, and had their soluble salt contents determined. The Karibib marble blocks showed observable deterioration, whilst the Damara granite ones were characterized by strength increases and pore-filling by soluble salts.

These studies using rock blocks have confirmed that highly aggressive ground conditions occur within and around coastal salt pans producing weathering 'hotspots', whereas inland conditions are less severe and salt weathering is limited to topographic hollows where suitable microclimatic conditions might prevail (Viles 2005). A potentially very important source of moisture for weathering processes in near-coastal areas is fog. Directly fog provides a source of moisture for a range of chemical and physical weathering processes, especially salt weathering, whilst indirectly fog can support diverse lichen communities which themselves have been shown to play a key role in weathering here (Viles and Goudie 2000).

As well as fog, dew may be an important source of moisture for weathering processes in the Namib Desert. Henschel and Seely (2008, p. 364) stated that dew in the Namib "occurs very frequently during most of the year". They report that in 2001 dew occurred on 53 nights at Gobabeb. As Eckardt et al. (2012) note it is hard to distinguish fog from dew and it can be highly patchy. As with fog, dew is a key additional source of moisture which can contribute to the weathering of rock surfaces and both directly through facilitation of chemical reactions and indirectly through its enhancement of lichen growth. Information on the amount and importance of dew and fog precipitation at Gobabeb and Kleinberg is provided by Kaseke et al. (2012).

The Namib also shows large diurnal swings in relative humidity, which are important in terms of salt crystallization and hydration cycles. Analyses by Viles (2005, Table 5) indicate that at Vogelfederberg there may be over 20 days a month when relative humidity values cross crucial thresholds for the hydration and crystallization of sodium sulphate (one of the common desert salts proven to be highly effective as a weathering agent). Given that, in reality, a number of salts in mixtures are often found on rocks in the Namib, (e.g. sodium chloride and calcium sulphate mixtures as reported by Viles and Goudie 2007), it is likely that key thresholds (which vary from salt to salt) will be crossed very frequently. If one considers the combined occurrence of rain, fog precipitation and dew fall it is apparent that at Gobabeb moisture is potentially deposited on rock surfaces on about 40 % of the days of the year (Henschel and Seely 2008, p. 364). Some rock surfaces will experience longer periods of wetness as a result of microenvironmental conditions (shading, topographic hollows etc.) which favour slow evaporation rates.

In addition to fog, rain and dew, another important source of moisture in some parts of the central Namib is groundwater discharge via springs, seeps and pans (Viles and Goudie 2013). This is particularly true of the coastal zone to the north of Swakopmund, where there are large areas of moist ground that are readily identifiable on satellite imagery. However, ground water discharge is also evident at some localities further inland, as, for example, on the pediment slopes on the southern side of Rössing Mountain, along the Swakop canyon, in the bed of the Soutrivier (Day and Seely 1988) and also at Ubib in the central Namib plains (Brain and Koste 1993). It is clear that groundwater is of great importance for weathering in the central Namib as it provides an often perennial source of moisture, and one which usually has high dissolved salt concentrations. Furthermore, groundwater discharge provides a deep seated and consistent source of moisture for weathering, rather than the more superficial and intermittent delivery of moisture from fogs, dews and rainfall.

References

Brain CK, Koste W (1993) Rotifers of the genus Proales from saline springs in the Namib Desert, with the description of a new species. Hydrobiologia 255(256):449–454

Bulley BG (1983) The engineering geology of Swakopmund. Geological Survey, report, Windhoek 13/4/2/90

Cockburn H, Seidl HA, Summerfield MA (1999) Quantifying denudation rates on inselbergs in the Central Namib Desert using in situ produced cosmogenic ^{10}Be and ^{26}Al. Geology 27:399–402

Cooke RU, Smalley IJ (1968) Salt weathering in deserts. Nature 220:1226–1227

Day JA, Seely MK (1988) Physical and chemical conditions in an hypersaline spring in the Namib Desert. Hydrobiologia 160:141–153

Eckardt FD, Soderberg K, Coop LJ, Muller AA, Vickery KJ, Grandi RD, Jack C, Kapalanga TS, Henschel J (2012) The nature of moisture at Gobabeb, in the central Namib Desert. J Arid Environ 93:7–19

Goudie AS, Parker AG (1998) Experimental simulation of rapid rock block disintegration by sodium chloride in a foggy coastal desert. J Arid Environ 40:347–355

Goudie AS, Viles HA (1997) Salt weathering hazards. Wiley, Chichester

Goudie AS, Viles HA, Parker AG (1997) Monitoring of rapid salt weathering in the Central Namib Desert, using limestone blocks. J Arid Environ 37:581–598

Henschel JR, Seely MK (2008) Ecophysiology of atmospheric moisture in the Namib Desert. Atmos Res 87:362–368

Kaseke KF, Mills AJ, Esler K, Henschel J, Seely MK, Brown R (2012) Spatial variation of "non-rainfall" water input and the effect of mechanical soil crusts on input and evaporation. Pure appl Geophys 169:2217–2229

Magee AW, Bull PA, Goudie AS (1988) Chemical textures on quartz grains: an experimental approach using salts. Earth Surf Proc Landf 13:665–676

Schiavon N, Chiavari G, Schiavon G, Fabbri D (1995) Nature and decay effects of urban soiling on granitic building stones. Sci Total Environ 167:87–101

Viles HA (2005) Microclimate and weathering in the Central Namib Desert, Namibia. Geomorphology 67:189–209

Viles HA, Goudie AS (2000) Weathering, geomorphology and climatic variability in the Central Namib Desert. In: McLaren SJ, Kniveton DR (eds) Linking climate change to land surface change. Kluwer, Dordrecht, pp 65–82

Viles HA, Goudie AS (2007) Rapid salt weathering in the coastal Namib desert: implications for landscape development. Geomorphology 85:49–62

Viles HA, Goudie AS (2013) Weathering in the central Namib Desert, Namibia: controls, processes and implications. J Arid Environ 93:20–29

The Namib Plains: Gypsum Crusts and Stone Pavements

14

Abstract

In the central Namib gypsum crusts (gypcrete) are a widespread surface material, particularly in drier areas. Their origin is complex. One model to explain the presence of sulphate in the coastal tracts of the central Namib is that marine biogenic hydrogen sulphide (H_2S), developed on the highly productive Namibian shelf, erupts from time to time and is carried inland by south westerly winds. Another suggested source of sulphate is anhydrite and bedrock sulphide. Some of the Namib gypsum may be distributed across the desert surface by dust storms that have deflated evaporite material from the many small pans and sabkhas identified on satellite images. Another important surface type in the Namib Plains is the stone or desert pavement. These are armoured surfaces composed of a mosaic of fragments, usually only one or two stones thick, set on or in matrices of finer material. They are formed by a range of processes that cause coarse particle concentration at the surface: (a) deflation of fine material by wind; (b) removal of fines by surface runoff and/or creep; and (c) processes causing upward migration of coarse particles to the surface.

14.1 Gypsum Crusts

A number of different crusts have formed in the central Namib Desert. The most widespread crust type in the coastal sector is the gypsum crust or gypcrete ($CaSO_4 \cdot 2H_2O$) (Watson 1979). It is generally found in areas where the rainfall is less than 50 mm per year and in a belt within 50–70 km of the coast between Walvis Bay and the Ugab River (Eckardt 1996; Eckardt and Spiro 1999; Bao et al. 2000, 2001; Eckardt et al. 2001). Petrographic details are provided by Watson (1985, 1988) and profile details by Wilkinson (1990) and Heine and Walter (1996a).

Watson (1985) suggested that there were three main types of gypsum crust. The first of these he termed 'bedded crusts', which he believed originated as laminated, shallow-water evaporites which accumulate when shallow pools or lagoons evaporate to dryness. The second type he termed 'subsurface crusts'. These, he believed, occurred in two forms, one made up of large crystals (>1 mm in diameter), often called desert roses, and the other of mesocrystalline (finer than 1.0 mm in diameter) material. The former tended to develop in low-lying situations in association with evaporation from high groundwater levels (the *per ascensum* type), while the latter, which are widespread in the Namib and often have columnar structures, had an illuvial or *per descensum* origin (i.e. they formed by the sub-surface accumulation of gypsum materials brought in from the surface, leached downwards, and then precipitated when the infiltrating water dries up). The third category he termed 'surface crusts'. These are the result of the exhumation and degradation of subsurface crusts.

Gypsum-rich materials may cover around 207 million ha of Earth's surface and the majority occur where the mean annual rainfall is less than 200–250 mm. This is because gypsum is semi-soluble ($\sim$2.6 g/l at 25 °C) and is normally leached out under higher rainfall conditions. Gypsum crusts are recorded in many of the world's deserts, but it is probably in Tunisia and the Namib that they show their greatest development (Watson 1979, 1988). The relative aridity of the Namib favours gypsum accumulation (Heine and Walter 1996b), but considerable debate has surrounded the source of the sulphate that makes up the crusts. Gypsum crusts are relatively less well developed in the southern part of the Namib Desert (the Sperrgebiet), and this may in part be due to extreme wind scour preventing crusts from accumulating (Miller 2008, Sect. 25.6.8) and in part to

A. Goudie and H. Viles, *Landscapes and Landforms of Namibia*, World Geomorphological Landscapes,
DOI 10.1007/978-94-017-8020-9_14, © Springer Science+Business Media Dordrecht 2015

the lack of sulphate as large sulphate inputs occur preferentially in the central Namib.

There are two main conceptual models to explain the source of sulphate for the Namib gypsum crusts (Figs. 14.1 and 14.2). The first model proposed by Henno Martin to explain the presence of sulphate in the coastal tracts of the central Namib is that marine biogenic hydrogen sulphide (H_2S), developed on the highly productive Namibian shelf, erupts from time to time (with malodorous consequences) and is carried into the desert by south westerly winds (Logan 1960; Martin 1963; Wilkinson et al. 1992) (Fig. 14.1). The location of the Central Namib favours such a scenario because it is both the foggiest part of the coast (Olivier 1995) and is adjacent to the thickest organic-rich sediment accumulations on the Namibian shelf, which are fed by the highly productive surface waters of the Benguela Current. Remote sensing studies suggest that such sulphide eruptions are large in extent, of frequent occurrence and of long duration (Brüchert et al. 2009). Some early analyses of fog water showed that it had high sulphate contents (Goudie 1972), although later analyses seemed to refute this. Another suggested source of the sulphate is anhydrite and bedrock sulphide (Cagle 1975). Some sulphate could be derived from deflation of inland saline pans (e.g. Etosha) and its transport to the western Namib by easterly 'berg' winds. However, sulphur isotope studies of potential sulphate sources and chemical analyses of fog water indicate that most fog water is extremely pure (Eckardt and Schemenauer 1998) and that neither biogenic H_2S nor bedrock are the source of the sulphate (Eckardt and Spiro 1999).

The second conceptual model to explain the source of sulphur for the Namib Desert gypsum crusts proposes that the primary source is the oxidation of marine dimethyl sulphide (DMS), with secondary reworking. This aerosol-derived material (Fig. 14.2) may have been accumulating in the old and hyper-arid Namib since the late Miocene and is redistributed by wind action (including deflation from numerous pans) and sporadic surface runoff (Eckardt et al. 2001). Evidence from $\Delta^{17}O$ studies by Bao et al. (2000, 2001) seems to point in the same direction.

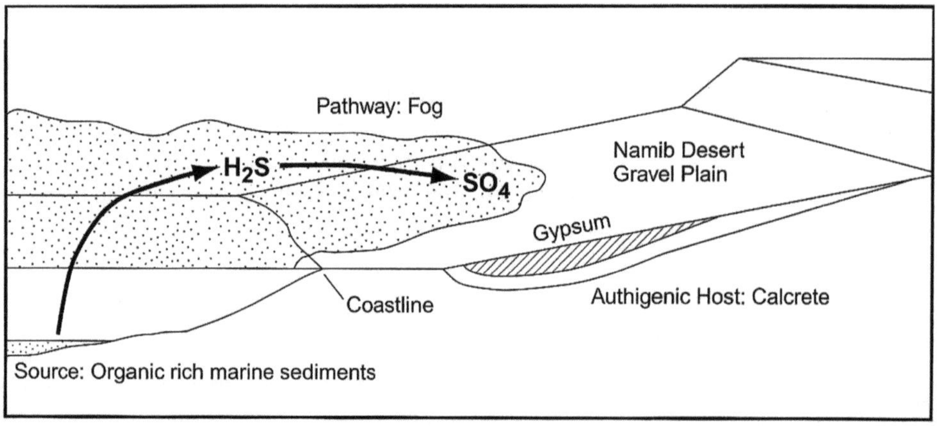

Fig. 14.1 Martin's model of gypsum crust development (courtesy of Dr. Frank Eckardt)

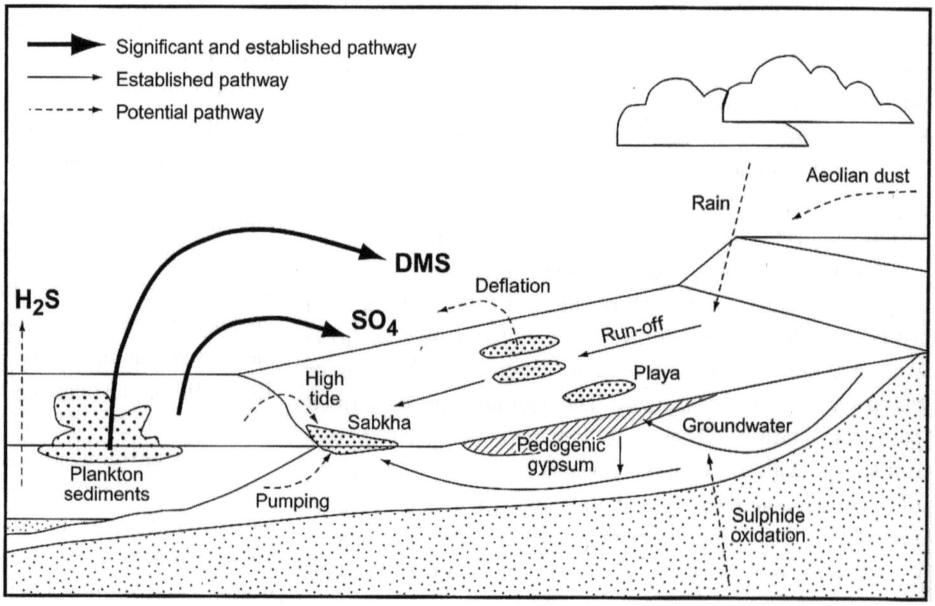

Fig. 14.2 Eckardt's model of gypsum crust development (courtesy of Dr. Frank Eckardt)

Whatever its primary source, some of the Namib gypsum may be distributed across the desert surface by dust storms that have deflated evaporite material from the many small pans and sabkhas that have been identified on satellite images (Eckardt et al. 2001). This gypsum contributes to further weathering through its contribution to salt weathering processes. Furthermore, the presence of gypcrete up to 1.5 m thick within soil profiles affects the balance between weathering and erosion across large swathes of the coastal gravel plains, probably contributing to the relatively low rates of long term denudation recorded there by cosmogenic isotope studies.

14.2 Stone Pavements

Another important surface type in the Namib Plains is the stone or desert pavement. These are armoured surfaces composed of a mosaic of fragments, usually only one or two stones thick, set on or in matrices of finer material comprising varying mixtures of sand, silt or clay.

They are formed by a range of processes that cause coarse particle concentration at the surface: (a) the classic mechanism of deflation of fine material by wind (Fig. 14.3); (b) removal of fines by surface runoff and/or creep; and (c) processes causing upward migration of coarse particles to the surface. In addition, it has become increasingly clear that pavements may evolve in close association not only with aeolian erosion but also with dust deposition and soil-profile differentiation caused by weathering.

Pavements have usually been explained as being produced by the classical mechanism of deflation of fine material from the surface, which leaves a residue, lag or armour of coarse particles. The concentration of coarse particles has been seen as a function of their distribution in the original sediment and the extent of deflation. However, lateral movement of fine materials could also be achieved by the second mechanism, i.e. removal by runoff or creep. Experimental observations show that some pavements are often composed, at least in part, of coarse particles that remain after finer materials have been dislodged and removed by raindrop erosion and running water (Wainwright et al. 1995). Plainly the role of sheetfloods should not be ignored as a horizontal transport mechanism (Williams and Zimbelman 1994; Dietze et al. 2013).

The third group of hypotheses involves vertical rather than horizontal movement of particles. The concentration of coarse particles at the surface and at depth, and the relative scarcity of coarse particles in the upper soil profile suggest that stones may have moved upwards through the soil to the surface by cycles of freezing and thawing, wetting and drying, or salt heave. Of these the most effective and widespread migration mechanism in deserts is thought to be associated with wetting and drying of the surface soil (McFadden et al. 1987). When a soil containing expanding clay minerals is wetted, it expands and a coarse particle is lifted slightly. As the soil shrinks on drying, cracks are produced around the particle and within the soil. Because of its large size the coarse particle cannot move down into the cracks, whereas finer particles can. The net effect is an upward displacement of the coarse particle. Subsidiary mechanisms of upward migration applicable to the Namib Desert may be salt heave (Searl and Rankin 1993), and the activity of soil fauna, including ants, termites, and burrowing mammals. Whether or not bioturbation causes stone pavement formation or disruption is, however, still a matter of debate. On the one hand, churning and burrowing may bring fine material to the surface, where it can be deflated, while on the other the process may cause coarse particles to sink and for homogenisation to occur. Under higher rainfall conditions, e.g. during pluvials, it is probable that pavement disruption predominates. Lichens may also contribute to the stabilisation of pavement surfaces once established, and there is good evidence that the lichen-dominated biological soil crusts along the coastal area of the central Namib Desert are agents of bioprotection (Lalley and Viles 2008; Viles 2008). Once damaged or destroyed by, for example, off-road driving, this bioprotection is lost and fine material is easily blown away. Finally, in recent years it has become appreciated that significant amounts of dust are delivered to desert

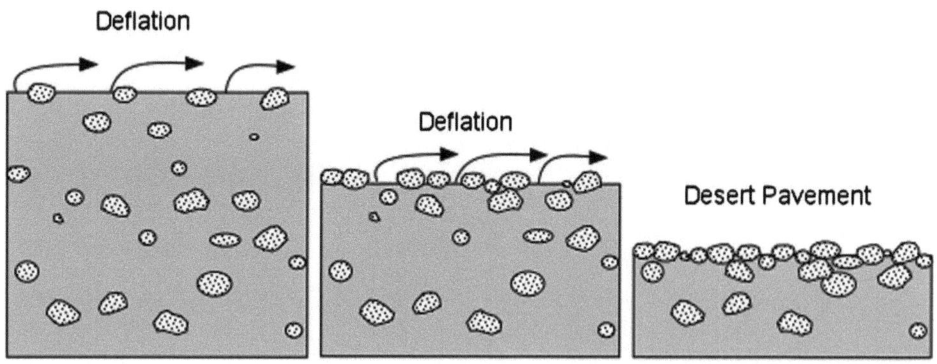

Fig. 14.3 The deflational model of stone pavement formation

surfaces by dust storms, and it is therefore inevitable that such dust contributes to the development of stone pavements.

References

Bao H, Thiemens MH, Farquar J, Campbell DA, Lee CCW, Heine K, Loope DB (2000) Anomalous ^{17}O compositions in massive sulphate deposits on the Earth. Nature 406:176–178

Bao H, Thiemens MK, Heine K (2001) Oxygen—17 excesses of the Central Namib gypcretes: spatial distribution. Earth Plan Sci Lett 192:125–135

Brüchert V, Currie B, Peard KR (2009) Hydrogen sulphide and methane emissions on the central Namibian shelf. Progr Oceanogr 83:169–179

Cagle FR (1975) Evaporite deposits of the Central Namib Desert. MSc Thesis, University of New Mexico, Albuquerque

Dietze M, Groth J, Kleber A (2013) Alignment of stone-pavement clasts by unconcentrated overland flow—implications of numerical and physical modelling. Earth Surf Proc Land 38:1234–1243

Eckardt F (1996) The distribution and origin of gypsum in the Central Namib Desert, Namibia, D.Phil thesis. University of Oxford

Eckardt F, Schemenauer RS (1998) Fog water chemistry in the Namib Desert, Namibia. Atmos Environ 32:2595–2599

Eckardt F, Drake N, Goudie AS, White K, Viles H (2001) The role of playas in pedogenic gypsum crust formation in the Central Namib Desert: a theoretical model. Earth Surf Proc Landf 26:1177–1193

Eckardt F, Spiro B (1999) The origin of sulphur in gypsum and dissolved sulphate in the Central Namib Desert, Namibia. Sed Geol 123:255–273

Goudie A (1972) Climate, weathering, crust formation, dunes and fluvial features, of the Central Namib Desert, near Gobabeb, South West Africa. Madoqua ser II 1:15–31

Heine K, Walter R (1996a) Gypcretes of the central Namib Desert, Namibia. Palaeoecol Afr 24:173–199

Heine K, Walter R (1996b) Die Gipkrustenböden der Zentralen Namib (Namibia) und ihr paläoklimatischer Aussagewert. Petermanns Geogr Mitt 140:237–253

Lalley JS, Viles HA (2008) Recovery of lichen-dominated crusts in a hyper-arid desert. Biodivers Conserv 17:1–20

Logan RF (1960) The Central Namib Desert, South West Africa. National Academy of Sciences/National Research Council Publication, 758, Washington DC

Martin H (1963) A suggested theory for the origin and a brief description of some gypsum deposits of South-West Africa. Trans Geol Soc S Afr 66:345–351

McFadden LD, Wells SG, Jercinovich MJ (1987) Influences of eolian and pedogenic processes on the origin and evolution of desert pavements. Geology 15:504–508

Miller RMcG (2008) The geology of Namibia, vol 3. Geological Survey of Namibia, Windhoek

Olivier J (1995) Spatial distribution of fog in the Namib. J Arid Environ 29:129–138

Searl A, Rankin S (1993) A preliminary petrographic study of the Chilean nitrates. Geol Mag 130:319–333

Viles HA (2008) Understanding dryland landscape dynamics: do biological crusts hold the key? Geogr Compass 2(3):899–919

Wainwright J, Parsons AJ, Abrahams AD (1995) A simulation study of the role of raindrop erosion in the formation of desert pavement. Earth Surf Proc Land 20:277–291

Watson A (1979) Gypsum crusts in deserts. J Arid Environ 2:3–20

Watson A (1985) Structure, chemistry and origin of gypsum crusts in Southern Tunisia and the Central Namib Desert. Sedimentology 32:855–875

Watson A (1988) Desert gypsum crusts as paleoenvironmental indicators: a micropetrographic study of crusts from Southern Tunisia and the Central Namib Desert. J Arid Environ 15:19–42

Wilkinson MJ (1990) Palaeoenvironments in the Namib Desert. The lower Tumas basin in the late Cainozoic. University of Chicago Research Papers p 231

Wilkinson M (1992) J, Hems DR, Whitehead VS (1992) Albedo patterns and gypsum generation in the Central Namib desert: land, sea and air interactions on an arid west coast. Geosci Remote Sens Symp 2:1565–1567

Williams SH, Zimbelman JR (1994) Desert pavement evolution: an example of the role of sheetflood. J Geol 102:243–248

15 The Homeb Silts

Abstract
The Homeb Silts are 26 m thick deposits which form some very clear flat-topped terraces and castle-shaped remnants, with very evident stratification, along the sides of the Kuiseb River. They occur at up to 45 m above the present channel floor. There have been four different hypotheses for their origin. The first is that they are lake sediments that arose when the Kuiseb was dammed by northward moving dunes coming out of the Namib Sand Sea. The second hypothesis is that they are river end-point sediments laid down by the Kuiseb under arid conditions when the river had less energy than today. The third hypothesis is that they are flood plain sediments laid down by an aggrading Kuiseb under semi-arid conditions that were wetter than today. The fourth hypothesis, which is not that different from the previous one, is that they are river flood sediments, called slackwater deposits, which were laid down by floods associated with intense precipitation events in the headwater regions. Also controversial is the age of the deposits, and different dating methods have given varying ages that span the late Holocene to the Late Pleistocene.

On the Kuiseb, upstream from Gobabeb, is the small Topnaar village of Homeb. It has given its name to one of the most striking and controversial landform sites in Namibia—the terraces of the Homeb Silts. These can be seen from the dirt road that descends from the Namib Plains into the Kuiseb Valley and then to the village (Fig. 15.1). These 26 m thick deposits form some very clear flat-topped terraces and castle-shaped remnants, with very evident stratification (Fig. 15.2). They occur at up to 45 m above the present channel floor. They have been greatly eroded and show some of the characteristics of a badland landscape. What has not been clear is their origin. As Miyamoto (2010) has pointed out, there have been four different hypotheses. The first of these is that the silt terrace deposits are lake sediments that arose when the Kuiseb was dammed by northward moving dunes coming out of the Namib Sand Sea (Goudie 1972; Rust and Wieneke 1980). However, as Heine and Heine (2002) have pointed out, the silt layers show neither the varved nor the deltaic stratification of lake sediments, and chemical precipitates are almost absent. The second hypothesis is that they are river end-point sediments laid down by the Kuiseb under arid conditions (Marker and Müller 1978) when the river had less energy than today. The third hypothesis is that they are flood plain sediments laid down by an aggrading Kuiseb under semi-arid conditions that were wetter than today. Despite the overall fine-grained texture of the Homeb sediments, there are also coarser beds with structures indicating rapid deposition from a shallow, fast-flowing, flash-flooding, sediment-laden stream (Ward 1987; Smith et al. 1993). The fourth hypothesis, which is not that different from the previous one, is that they are river flood sediments, called slackwater deposits, which were laid down by floods associated with intense precipitation events in the headwater regions (Heine and Heine 2002; Heine et al. 1999; Miyamoto 2010). They accumulated in the low energy setting of tributary valleys and were laid down in a succession of floods.

Also controversial is the age of the deposits, and different dating methods have given varying ages that span the late Holocene to the Late Pleistocene. Miyamoto (2010), using radio-carbon dating, believed that almost all the silts were deposited between 25 and 19 thousand year ago, and Heine and Heine (2002) report broadly similar radiocarbon dates equating with the Last Glacial Maximum. Conversely, optically stimulated luminescence (OSL) dates obtained by Bourke et al. (2003) gave ages between 6.3 and 9.8 ka, i.e.

A. Goudie and H. Viles, *Landscapes and Landforms of Namibia*, World Geomorphological Landscapes,
DOI 10.1007/978-94-017-8020-9_15, © Springer Science+Business Media Dordrecht 2015

Fig. 15.1 The Homeb Silts

Fig. 15.2 The Homeb Silts

during the early to mid Holocene. More recent OSL dates from Srivastava et al. (2006) indicate that rapid deposition of the Homeb Silts occurred at around 15 ka and 6 k during climate transitions from arid to humid. They argued that the transition to a wetter climate favoured channel aggradation because that is when the stream had enough energy to transport sediment and yet there was no vegetation to prevent its removal from hillslopes. As the climate becomes wetter, the vegetation cover became greater which promoted an increase in the water/sediment ratio and the incision of previously deposited sediments. Evidently, there is need for further dating and palaeoenvironmental work to clarify the age and modes of formation of these intriguing silt deposits and to compare them with the Amspoort silts and Clay Castles in similar ephemeral catchments further north (see Chap. 8).

References

Bourke MC, Child A, Stokes S (2003) Optical age estimates for hyper-arid fluvial deposits at Homeb, Namibia. Quat Sci Rev 22:1099–1103

Goudie A (1972) Climate, weathering, crust formation, dunes and fluvial features, of the Central Namib Desert, near Gobabeb, South West Africa. Madoqua Series II 1:15–31

Heine K, Heine JT (2002) A paleohydrologic reinterpretation of the Homeb Silts, Kuiseb River, central Namib Desert (Namibia) and palaeoclimatic implications. Catena 48:107–130

Heine K, Heine C, Kühn T (1999) Slackwater deposits der Namib-Wüste (Namibia) und ihr paläoklimatischer Aussagewert. Zbl Geol Palaont Teil 1(5/6):587–613

Marker ME, Müller D (1978) Relict vlei silt of the middle Kuiseb Valley, South West Africa. Modoqua 11:151–162

Miyamoto S (2010) Late Pleistocene sedimentary environment of the "Homeb Silts" deposits, along the middle Kuiseb River in the Namib Desert, Namibia. Afr Study Monographs Suppl 40:51–66

Rust U, Wieneke F (1980) A reinvestigation of some aspects of the evolution of the Kuiseb River valley upstream of Gobabeb. Madoqua 12:163–173

Smith RMH, Mason TR, Ward JD (1993) Flash-flood sediments and ichnofacies of the Late Pleistocene Homeb Silts, Kuiseb River, Namibia. Sed Geol 85:579–599

Srivastava P, Brook GA, Marais E, Morthekai P, Singhvi AK (2006) Depositional environment and OSL chronology of the Homeb silt deposits, Kuiseb River, Namibia. Quat Res 65:478–491

Ward JD (1987) The Cenozoic succession in the Kuiseb Valley, Central Namib Desert. Geol Surv South West Africa/Namibia Memoir 9

Calcretes: The Kamberg Calcrete Formation and the Karpencliff Conglomerate

16

Abstract

Calcretes are calcium carbonate-rich duricrusts and are extensively developed in Namibia. Some of the most beautiful calcretes are those that cap the Karpencliff Conglomerate on the margins of the Kuiseb canyon. These have been termed the Kamberg Calcrete Formation. The calcrete pre-dates the canyon incision of the Kuiseb and so may date towards the end of the Miocene. The calcrete formed under semi-arid conditions over a long period of relative landscape stability, perhaps half a million years in length. The chapter discusses the various hypotheses that have been put forward to account for calcrete formation.

To the west of the 500 mm isohyet line a carbonate-rich type of duricrust, called calcrete, is almost ubiquitous in southern Africa and is encountered in river sections, around lake shorelines, and in countless borrow-pits along roads. In particular it is dominant in a great arc running down through the interior of Namibia into the Northern Cape of South Africa. Borehole records indicate that complex calcrete profiles can attain considerable thickness—sometimes over 30 m—most notably in a belt associated with the Auob, Nossob, Molopo and Kuruman rivers (Goudie 1973). Even thicker calcrete accumulations occur in the Ovambo Basin, where the Etosha Calcrete may be up to 120 m thick (see Chap. 6).

Although gypcretes dominate the drier, western portions of the Namib (see Chap. 14), there are some extensive spreads of calcrete here as well. They cap, for example, some of the Kuiseb terraces near Gobabeb (Lancaster et al. 2000; Yamagata and Mizuno 2005) and they also cement the pediments around Spitzkoppe and also in the vicinity of Erongo (Blümel and Vogt 1979; Blümel 1981). Some of the Namibian calcretes are of Lower Pliocene or Miocene age (Eitel 1994; Blümel and Eitel 1994) and tend to be rich in palygorskite (Eitel 2000). Calcretes in ancient palaeodrainage channels can be hosts for uranium, as at Langer Heinrich (Trittschack and Borg 2008). Elsewhere in Namibia calcrete caps the Weissrand Plateau (see Chap. 22), the Grunau area in the south of the country, and the terraces along the Auob and Nossob Rivers. The main outcrops of the KambergCalcrete are notable for the small solutional depressions—dayas—that dot their surfaces (see also Chap. 22). These can be seen on the surface to the south of the Kuiseb between Donkerhoek in the north and Spaarwater in the south.

Some of the most beautiful calcretes in Namibia are those that cap the Karpencliff Conglomerate on the margins of the Kuiseb canyon (Fig. 16.1). These have been termed the Kamberg Calcrete Formation (Ward 1987). The Karpencliff Conglomerate overlies the Tsondab Sandstone and was deposited in a proto-Kuiseb and a proto-Gaub Valley. Some facies indicate deposition in an alluvial fan environment. They then became cemented, with profiles up to 5 m thick developing. The calcrete pre-dates the canyon incision of the Kuiseb and so may date towards the end of the Miocene. The calcrete formed under semi-arid conditions over a long period of relative landscape stability, perhaps half a million years in length. Netterberg (1967) developed a classification of carbonate enrichment types, which included calcified soils, powder, nodular, honeycomb, hardpan, laminar and

A. Goudie and H. Viles, *Landscapes and Landforms of Namibia*, World Geomorphological Landscapes,
DOI 10.1007/978-94-017-8020-9_16,

Fig. 16.1 The calcrete-capped Karpencliff Conglomerate above the Kuiseb Canyon

boulder calcretes. This useful scheme sees calcified soils as the least developed form. With progressive calcification a powder calcrete develops, which is characterised by loose carbonate silt or sand particles with few or no host soil particles or nodules present. Nodular (or glaebular) calcrete is the next stage. As calcrete nodules increase in size and number they coalesce to form a honeycomb calcrete, the voids in which are still filed by soil material. Next, when the voids of the honeycomb calcrete become infilled or cemented, a hardpan calcrete develops. Some of these are indurated layers up to 3 m thick. These may often be capped with a laminar horizon. Finally, the calcrete may become degraded, producing a boulder calcrete or brecciated masses. Each of these types of calcrete can be observed in Namibia within various outcrops and exposures.

Calcretes can form in a wide variety of ways. Most calcretes are pedogenic (i.e. soil phenomena) though by no means all (Nash and McLaren 2003), and the thick Etosha Calcrete appears to have been precipitated from groundwater (Miller 2008, Chap. 24). In the case of pedogenic calcretes, such as those of the Weissrand Plateau, much calcium comes from dust or rain. This calcium is mobilized in surface horizons with high CO_2 contents, through which rainwater infiltrates, and deposited lower down the profile where CO_2 levels are less or from which water is lost by evaporation. This is the *per descensum* hypothesis (Fig. 16.2a). Until the late 1970s most models of calcrete formation were physiochemical.

Since then the role of organic processes has become clearer (Goudie 1996). Field relations and characteristics show that many calcretes are non-pedogenic, having been deposited near a water table (the *per ascensum* hypothesis) (Fig. 16.2b). The main mechanism may be evaporation from the capillary fringe, but carbonate deposition can be induced below a water table by the changing CO_2 contents of the water. Water-table calcretes develop especially in alluvium, where water-tables are near the surface and where the matrix is coarse. Other calcretes form by recementation by soil water of calcrete debris on pediment slopes (the detrital model) (Fig. 16.2c).

As with gypcrete (Chap. 14), at least some of the solutes which contribute to the lithification of calcrete come from weathering of carbonate rocks, such as marble. Over the long term, such calcretes also contribute to reduction of denudation rates by providing a resistant barrier to erosion. This effect might be one cause of the relatively low long—term denudation rates found from some areas of the Namib Desert in cosmogenic dating studies.

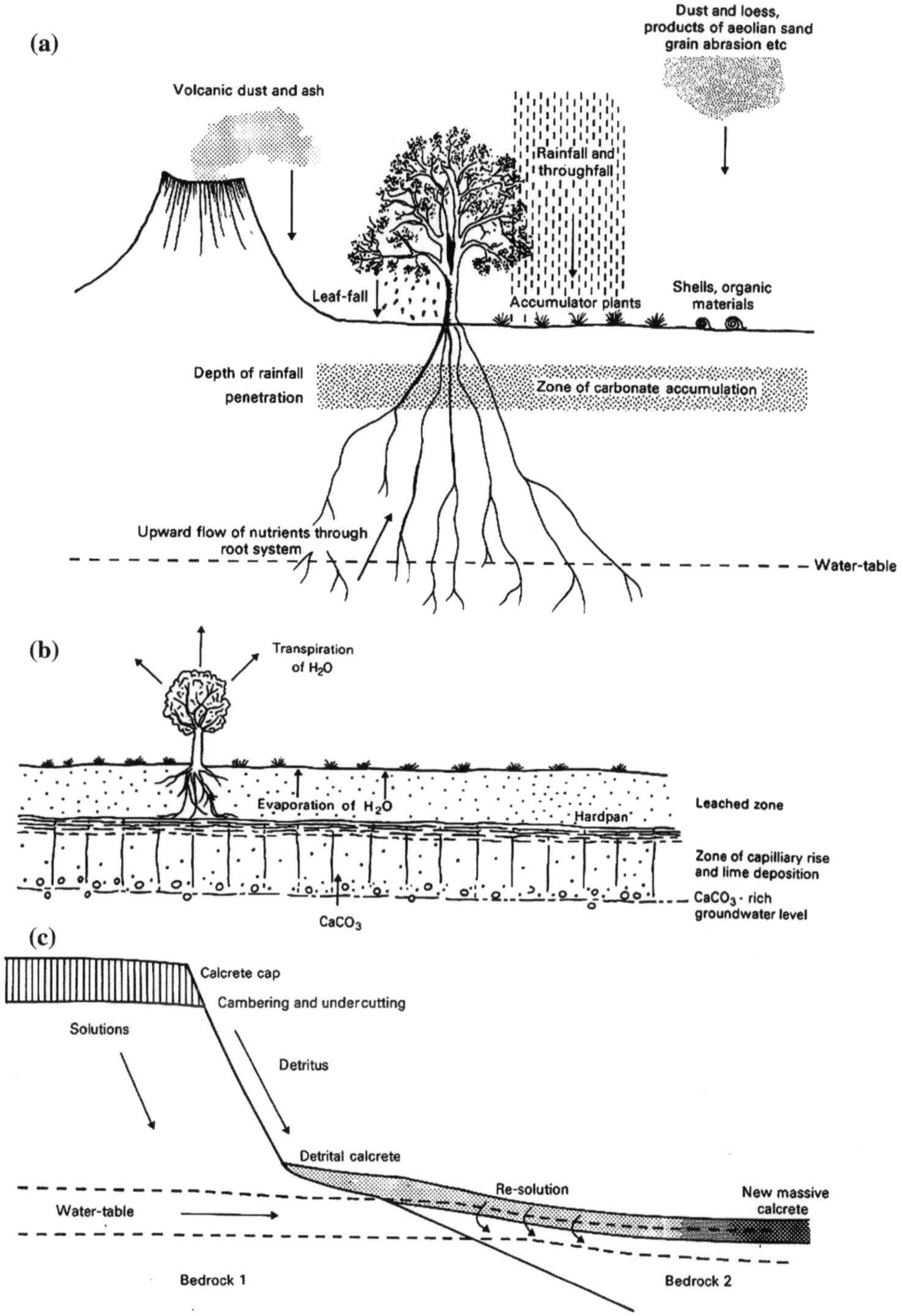

Fig. 16.2 **a** The per descensum model of calcrete development. **b** The per ascensum model of calcrete development. **c** The detrital model of calcrete development

References

Blümel WD (1981). Pedologische and Geomorphologische Aspekte der Kalkkrustenbildung in Südwestafrika und Südostspanien. Karlsruhe Geogr Hefte 10

Blümel WD, Vogt T (1979) Croûtes calcaires de Namibie. Rech Géogr à Strasbourg 12:55–64

Blümel WD, Eitel B (1994) Tertiäre Deckschichten und Kalkrusten in Namibia: Entstehung und geomorphologische Bedeutung. Z Geomorphol 38:385–403

Eitel B (1994) Kalkreiche Decksedimente und Kalkrustengenerationen in Namibia: Zur Frage der Herkunft und Mobilisierung des Calciumcarbonats. Stuttgarter Geogr Studien p 123

Eitel B (2000) Different amounts of pedogenic palygorskite in South West African Cenozoic calcretes: geomorphological, paleoclimatical and methodological implications. Z Geomorphol Suppl 121:139–149

Goudie AS (1973) Duricrusts in tropical and subtropical landscapes. Oxford University Press, Oxford

Goudie AS (1996) Organic agency in calcrete development. J Arid Environ 32:103–110

Lancaster N, Schaber GG, Teller JT (2000) Orbital radar studies of paleodrainages in the Central Namib Desert. Remote Sens Environ 71:216–225

Miller RMcG (2008) The Geology of Namibia. vol 3. Geological Survey of Namibia, Windhoek

Nash DJ, Mclaren SJ (2003) Kalahari valley calcretes: their nature, origins and environmental significance. Quat Int 111:3–22

Netterberg F (1967) Some roadmaking properties of South African calcretes. In: Proceedings of 4th Regional Conference on Soil Mechanisms and Foundation Engineering, pp 77–81

Trittschack R, Borg G (2008) Lithostratigraphie und Sedimentologie der Uranführended Langer Heinrich Formation/Namibia. Zentralblatt Geol Paläontol Teil 1, vol.Jg.2008. Heft 1:139–156

Ward JD (1987) The Cenozoic succession in the Kuiseb Valley. Central Namib Desert, Geol Surv South West Africa/Namibia Memoir 9

Yamagata K, Mizuno K (2005) Landform development along the middle course of the Kuiseb River in the Namib Desert, Namibia. Afr Study Monogr Suppl 30:15–25

17 Barchan Dunes: Walvis Bay and on the Skeleton Coast

Abstract

Barchans are mobile crescentic dunes that form transverse to the wind. Barchans have a range of forms from slim to pudgy and are found in areas with limited sand supply and narrowly bimodal winds. Near Walvis Bay there are particularly good examples, formed by winds coming from the south west. Long term studies, using air photographs, show that these dunes can move at rates of some metres per year. Large sand ripples are also found on some of the coastal dunes between Walvis Bay and Swakopmund.

Barchans are individual mobile dunes of crescentic shape, the two horns of which face in the direction of dune movement. Sand avalanching takes place on their lee sides. They are generally regarded as occurring in areas of limited sand supply, on planar surfaces, with a low precipitation (usually less than 100 mm per annum) and vegetation cover, and where winds are narrowly bimodal in direction (with a directional index that is normally around 0.7–0.9). At a global scale they are quantitatively of limited significance—less than 1 % of all dune sand on Earth is contained within them—but they can be locally dominant, as is the case in some parts of the coastal Namib Desert. They are variable in size, ranging in height from a few metres to over 500 m in the case of megabarchanoids (Bishop 2010).

Bourke and Goudie (2009) compared the shapes of barchans in the Namib and on Mars and developed the following classification scheme:

Classic symmetrical barchans—slim. The simplest form of barchan is the classic individual crescentic feature. Some of these are elegantly slim as shown by examples on the rocky plains to the south and east of Lüderitz and Elizabeth Bay (Fig. 17.1). They also appear to be rather angular in plan. They display a wide range of sizes, with some having widths as great as 500–600 m, and some being only a few tens of metres in width. The slim symmetrical type of barchans is a feature of areas with unidirectional winds and with low sand influx and high values for shear velocity (Parteli et al. 2007).

Classic symmetrical barchans—pudgy and fat. Some simple crescentic forms possess a larger area in relation to their width than the examples given above. The horns are relatively small in relation to the total mass of the dune and may be nearly absent. Such dunes have shapes reminiscent of kidneys, broad beans and pectens. Fat dunes occur in areas where there is a substantial sand influx and lower shear velocities (Parteli et al. 2007). Many of the world's barchans described in the literature appear to be fat rather than slim, and this is the shape of many of the barchans in the Kuiseb delta area (Barnes, 2001).

Classic symmetrical barchans—large, fat and unstable. Some barchans are large features, which may be termed mega-barchans. Over 500 m in width, they often have secondary features on their flanks, which may be indicative of instability. They may also shed small barchans onto the desert plains downwind. This appears to be an example of what Elbelrhiti et al. (2005) describe as 'surface-wave-induced instability'. They argue that dune collisions and changes in wind direction destabilize larger dunes and generate surface waves on their lee flanks. The resulting surface waves propagate at a higher speed than the dunes themselves, producing a series of small, new born barchans by breaking the horns of large dunes. This type of barchans can be seen near the Walvis Bay salt works.

Classic symmetrical barchans composed of smaller barchans. In southern Namibia a single classic barchan form some 400 m across and 700 m long that is predominantly

A. Goudie and H. Viles, *Landscapes and Landforms of Namibia*, World Geomorphological Landscapes,
DOI 10.1007/978-94-017-8020-9_17, © Springer Science+Business Media Dordrecht 2015

Fig. 17.1 Google Earth image of slim barchans near Elizabeth Bay. Scale bar 0.5 km. (© 2012 Google Image, Digital Globe)

made up of a cluster of smaller barchans exists. It may be an extreme an example of a proto-megabarchan (Cooke et al. 1993, p. 327).

Barchans developing into linear dunes. Following on from the classic model of Bagnold (1941) it is evident that some simple crescentic forms are deformed into linear (seif) features when they move into areas with changing wind regimes. Linear ridges some km long can develop downwind from the original barchans, creating a tadpole shape. Good examples of this can be found in northern Namibia along the Skeleton Coast (Fig. 17.2).

Barchan dunes developing into transverse ridges. There are many examples of classic individual barchans merging together with their neighbours to form ridges transverse to the formative winds. The original barchanoid and linguoid elements are clearly visible. It is generally believed that sand availability is a crucial control, and that with greater sand supply transverse dune ridges rather than individual isolated barchans will occur (see below).

Barchan convoys developing into linear ridges. Some intriguing linear dune ridges appear to be formed by convoys of approximately equally sized barchans. Wang et al. (2004) proposed this style of barchans merging in their model of complex linear dune formation.

Another type seems to have formed downwind of major nebkha fields. These develop from sand that has accumulated around bushes, rather than through the normal style of evolution from a non-anchored sand pile.

Moving barchans dunes can encroach upon houses, railways, roads and other types of infrastructure (Fig. 17.3). In the vicinity of Walvis Bay there are migrating barchan dunes that are so mobile that they can pose problems for roads, the railway and houses. They are driven primarily by winds coming from the south west (Fig. 17.4). These migrating dunes are of the classic symmetrical—large, fat and unstable type (Fig. 17.5). Good examples can be seen near the Radio Station and also at the Salt Works to the south of the Walvis Bay lagoon. The barchans near the Radio Station, illustrated in Figs. 17.6 and 17.7, have changed form with changes in wind regime and through time. Thus in July 2004, some of the barchans have plainly been moulded by northeasterly winds, as their arms face to the south west, whereas in September 2010 they point in the opposite direction. By 2010, the five dunes being shed from the westerly arm have disappeared, while those being shed from the easterly arm have migrated northwards and have developed a classic form.

The Salt Works barchans have a mean height of around 8 m, though some are nearly 20 m in height. The movement history of these latter dunes has been described over a period of some decades by Slattery (1990) and Barnes (2001) (Fig. 17.8). Slattery (1990) found a mean annual rate of movement of 13.5 m. Barnes, on the other hand, found that rates varied from decade to decade and year to year, ranging from just 4.2 m per annum between 1976 and 1988, to 56.1 m per annum between 1997 and 1999. She related these differences to changes in wind velocities during the period of observations.

One of the reasons why the Walvis Bay barchans are of particular interest is because unlike inland dunes they have

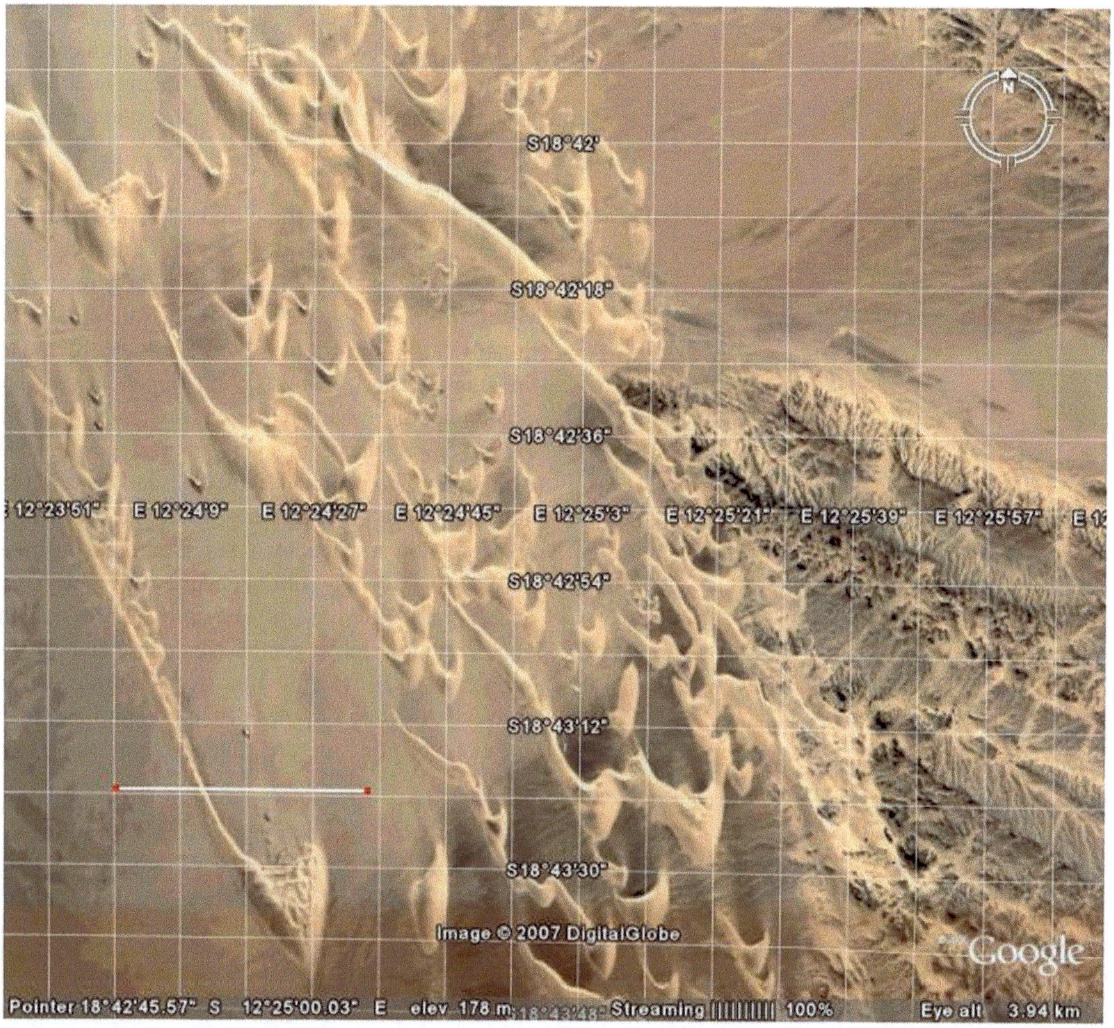

Fig. 17.2 Google Earth image of barchans elongating into linear dunes on the Skeleton Coast. Scale bar 1,000 m. (© 2007 Digital Globe)

Fig. 17.3 Dune encroaching upon the abandoned mining town of Kolmanskop, southern Namib

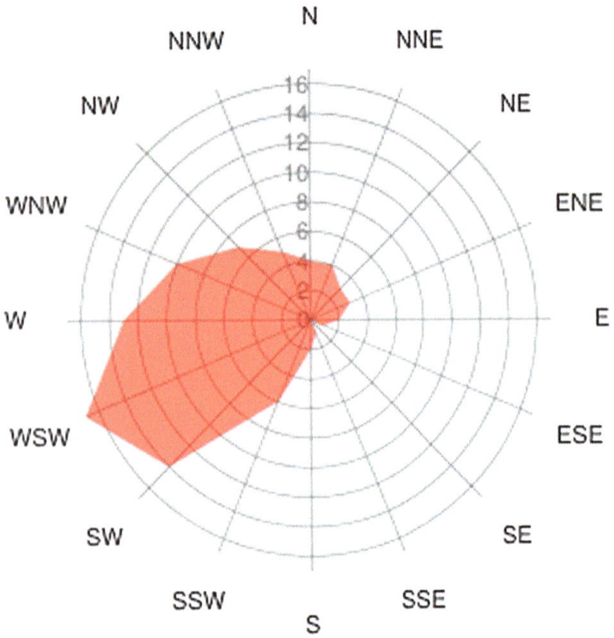

Fig. 17.4 Wind rose for Walvis Bay Airport for the whole year (http://www.windfinder.com/windstats/windstatistic_walvis_bay_airport.htm) (accessed 30th January 2014)

an appreciable salt content (Besler 1981) and are frequently wetted by fog, and this affects sand movement and some of the micro-forms developed upon them, including slab avalanches. They have recently been studied as a possible analogue for dunes on Mars.

Between Walvis Bay and Swakopmund the narrow coastal dune range is composed of a variety of dune types, but many of them are ridges that run transverse to the south westerly winds. On their seaward side there are large areas, visible from the main coastal road, where there are exceptionally large sand ripples, which are also perpendicular to the south westerly winds. Wind ripples are the smallest of aeolian bedforms and are present on almost all sand surfaces except those undergoing very rapid deposition. They generally trend perpendicular to the sand-transporting winds, although on sloping surfaces where the downwind component of grain movement is supplemented by gravity, they may be slightly flow oblique. Typically they have a wavelength of 13–300 mm and an amplitude of 0.6–14 mm. Like dunes, ripples have gentle windward slopes (in general between 8 and 13°) and rather steeper lee slopes (up to 30°). However, the large ripples of the coastal tract, many with a wavelength of c 3 m, are called granule ripples (Fryberger

Fig. 17.5 Barchans near Walvis Bay, 2011. These are the same ones as shown on the Google Earth Images. The large dune from which the smaller dunes have been shed is in the background

Fig. 17.6 Google Earth image of Walvis Bay barchans, July, 2004. Scale bar 200 m. (© 2012 Google Image, Digital Globe)

Fig. 17.7 Google Earth image of Walvis Bay barchans, September 2010. Scale bar 200 m. (© 2012 Google Image, GeoEye)

et al. 1992). These are 'aeolian bedforms comprised of a sandy core that is covered by a surface layer of granules, particles that are typically 1–2 mm in diameter' (Zimbelman et al. 2009). They tend to be significantly larger than wind ripples formed in well-sorted fine sand. They are also known as gravel ripples or megaripples (Isenberg et al. 2011).

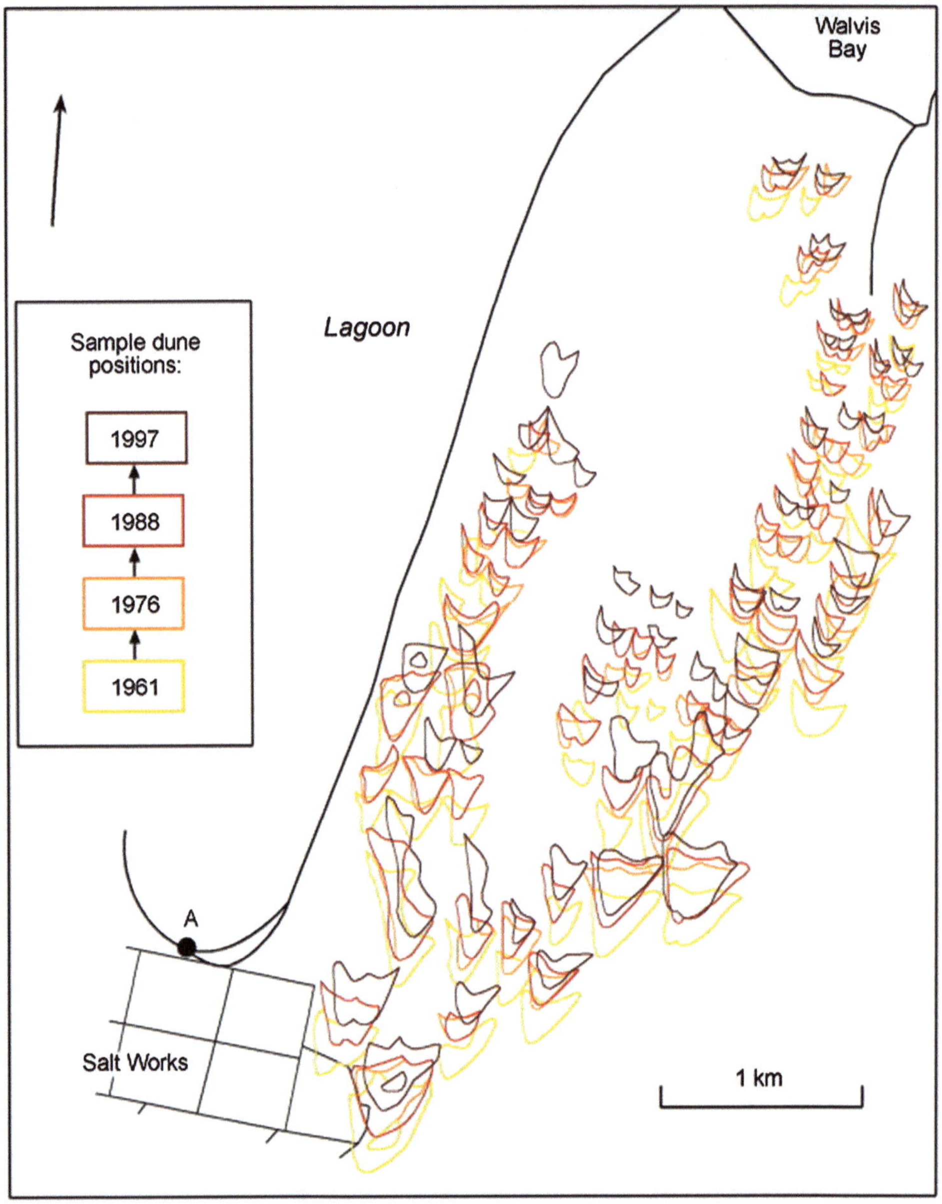

Fig. 17.8 The Walvis Bay barchans from 1961–1997 (modified after Barnes 2001, in Livingstone, 2013, Fig. 6)

References

Bagnold RA (1941) The physics of blown sand and Desert Dunes. Methuen, London

Barnes J (2001) Barchan dunes on the Kuiseb River delta, Namibia. S Afr Geogr J 83:283–292

Besler H (1981) Surface structures on Namib dunes caused by moisture. Namib und Meer 9:11–17

Bishop MA (2010) Nearest neighbour analysis of mega-barchanoid dunes, Ar Rub' al Khali, sand sea: The application of geographical indices to the understanding of dune field self-organization, maturity and environmental change. Geomorphology 120:186–194

Bourke MC, Goudie AS (2009) Varieties of barchan form in the Namib Desert and on Mars. Aeolian Res 1:45–54

Cooke RU, Warren A, Goudie A (1993) Desert geomorphology. UCL Press, London

Elbelrhiti H, Claudin P, Andreotti B (2005) Field evidence for surface-wave-induced instability of sand dunes. Nature 437:720–723

Fryberger SG, Hesp P, Hastings K (1992) Aeolian granule ripple deposits, Namibia. Sedimentology 39:319–331

Isenberg O, Yizhaq H, Tsoar H, Wenkart R, Karnieli A, Kok JF, Katra I (2011) Megaripple flattening due to strong winds. Geomorphology 131:69–84

Parteli EJR, Durán O, Herrmann HJ (2007) Minimal size of a barchan dune. Phys Rev E 75: article no. 011301

Slattery MJ (1990) Barchan migration on the Kuiseb River Delta, Namibia. S Afr Geogr J 72:5–10

Wang X, Dong Z, Zhang J, Qu J (2004) Formation of the complex linear dunes in the central Taklimakan Sand Sea, China. Earth Surf Proc Land 29:677–686

Zimbelman JR, Irwin RP, Williams SH, Bunch F, Valdez A, Stevens S (2009) The rate of granule ripple movement on Earth and Mars. Icarus 203:71–76

Namib Sand Sea

18

Abstract

The modern Namib Sand Sea is underlain by the Tsondab Sandstone Formation. Dating from c 21 million years ago, it is a red brown sandstone, up to 220 m thick. Large parts of the Tsondab Sandstone are aeolian though there are also pan and fluvial facies. Overlying the Tsondab Sandstone Formation between Lüderitz and the Kuiseb River is the modern Namib Sand Sea. At the coast crescentic dunes are dominant, including highly mobile barchans. The heart of the sand sea is dominated by linear dunes that are associated with more bi-directional wind regimes (SSW–SW and NE–E). Star dunes, characterised by their pyramidal morphology and radiating sinuous arms, are associated with complex, multidirectional wind regimes (SW–WSW, NE–E and N) that occur along the sand sea's eastern margin, close to the Great Escarpment. The sharply defined northern margin of the Sand Sea is formed by the Kuiseb River. Much of the sand may have been supplied by the Orange River to the coastal zone and then been blown inland. The colour of the dune sand of the Namib shows clear spatial trends, the reasons for which are discussed. In coastal areas dominated by crescentic dunes the sand is yellowish brown to light yellowish brown, whereas in eastern areas it becomes a very striking yellowish red.

18.1 Introduction

Dunes are one of the world's most fascinating and aesthetically pleasing landform types. As Ralph Bagnold remarked in his classic 'Physics of Blown sand and Desert Dunes' (1941, p. xxi):

> Here, instead of finding chaos and disorder, the observer never fails to be amazed by a simplicity of form, an exactitude of repetition and a geometric order unknown in nature on a scale larger than that of crystalline structure. In places vast accumulations of sand weighing millions of tons move inexorably, in regular formation, over the surface of the country, growing, retaining their shape, even breeding, in a manner which, by its grotesque imitation of life, is vaguely disturbing to an imaginative mind. Elsewhere the dunes are cut to another pattern—lined up in parallel ranges, peak following peak in regular succession like the teeth of a monstrous saw for scores, even hundreds of miles, without a break and without a change in direction, over a landscape so flat that their formation cannot be influenced by any local geographical features.

Under strong wind conditions, sand grains are transported across desert surfaces. When the wind velocity exceeds the threshold velocity that is required to initiate sand grain movement, the grains begin to roll along the ground, but after a short distance this gives rise to a bounding or jumping action called saltation. Grains are taken up a small distance into the airstream and then fall back to the ground in a fairly flat trajectory. The descending grains dislodge further particles and thereby the process of saltation is maintained across the surface. Dunes form because saltating grains tend to accumulate preferentially on sand-covered areas rather than on adjoining sand-free surfaces. This means that a small sand accumulation gets bigger and turns into a dune under some positive feedback conditions. The precise form of such a dune will depend on such factors as the wind regime, the degree of vegetation cover, the amount of sand available, and the size of the grains involved.

The Namib desert has a number of sand seas (*ergs*) in which these dunes occur—in southern Angola is the Curosa-Bahia dos Tigres sand sea, in northern Namibia are the Kunene (see Chap. 5) and Skeleton Coast sand seas, while in the south is the largest of them all—the Namib Sand Sea itself (Stone 2013). This magnificent area was inscribed as a UNESCO World Heritage Site in June 2013. The Namib Sand Sea is in many parts active today, but the amount of

A. Goudie and H. Viles, *Landscapes and Landforms of Namibia*, World Geomorphological Landscapes,
DOI 10.1007/978-94-017-8020-9_18, © Springer Science+Business Media Dordrecht 2015

sand present overall indicates that more than a million years may have been required for it to accumulate (Vermeesch et al. 2010; Garzanti et al. 2012).

18.2 The Tsondab Sandstone—Predecessor of the Namib Sand Sea

The modern Namib Sand Sea (Fig. 18.1) is underlain by what has been called the Namib Sandstone (Besler and Marker 1979) but which is now generally known as the Tsondab Sandstone Formation (Ward 1988). This is a truly remarkable phenomenon. Dating back to c 21 million years ago, it is a red brown sandstone, up to 220 m thick, and is especially well exposed to the south of Solitaire on the Conas Cliffs located on the farm Dieprivier. Large parts of the Tsondab Sandstone are aeolian (Kocurek et al. 1999), though there are also pan and fluvial facies. The structures in the aeolian deposits appear to have formed under a broadly similar wind regime to that of the present, with barchans in the west and linear and star forms in the east. They also contain the fossilized tracks and burrows of termites and golden moles that live in the Namib today, together with lithified ostrich eggs, spider webs, carnivore footprints, coprolites, and the remains of an extinct species of hyaena, *Crocuta dietrichi* (Senut et al. 1998; Pickford 2000; Ségalen et al. 2002; Morales et al. 2011). In effect the Tsondab Sandstone is a massive fossilized sand sea that at about 16 million years ago extended from the Orange to the Kunene, but much of which has now been eroded. The portion to the south of the Aus to Lüderitz road is called the Rooilepel Sandstone Formation (Miller 2008, pp. 25–22).

18.3 Dune Diversity

Overlying the Tsondab Sandstone Formation between Lüderitz and the Kuiseb River, the modern Namib Sand Sea (Lancaster 1989) covers some 34,000 km^2. Its main dune types have been mapped and described by Livingstone et al. (2010), Bullard et al. (2011) and Livingstone (2012) (Fig. 18.2), and their internal structures have been studied by McKee (1982).

At the coast crescentic dunes are dominant, including highly mobile barchans (Slattery 1990; Barnes 2001). Their horns point in the direction of movement, they have steep slopes (c 32°) on their lee sides and gentler slopes (2–10°) on their windward (stoss) sides, they have an ellipsoidal shape in plan-view, and have formed in response to the strong unidirectional (SSW) wind regimes that are prevalent in the coastal zone (Fig. 18.3). These can be seen inland from Walvis Bay (see Chap. 17) and in the vicinity of

Fig. 18.1 The location of the Namib Sand Sea (from Livingstone et al. 2010, Fig. 1)

Lüderitz, where they often cross the railway and the main road respectively. They have encroached on the old mining town of Kolmanskop. Where there is a large sand supply, the barchans sometimes coalesce to create transverse ridges. Elsewhere they may have the form of mega-barchans, with small features developed on a much larger crescentic base.

Also in the coastal zone are nebkhas. This is an Arabic term given to mounds of wind-borne sediment (sand, silt of pelletized clay) that have accumulated to a height of some metres around shrubs or other types of vegetation. Plants that gain their water supply from high groundwater levels (e.g. tamarisk) may often form the core around which accumulation occurs. The largest nebkhas (mega-nebkhas) accumulate around clumps of trees. Nebkhas are sometimes called shrub-coppice dunes (see Rango et al. 2000). They may occur on bigger dunes, in inter-dune areas, on pan surfaces, near wadis and on or behind beaches and sabkhas. Some nebkhas are more or less circular mounds, whereas others show clear elongation and consist of a long plume to the lee of the anchoring vegetation. Extensive fields of nebkhas occur just to the north of Lüderitz at Agate Bay and also behind Hottentots Bay.

The heart of the sand sea is dominated by linear dunes (Fig. 18.4) that are associated with more bi-directional wind regimes (SSW-SW and NE-E) (Bubenzer and Bolten 2008). The former winds blow inland from the South Atlantic Ocean and the latter sweep down the Great Escarpment from

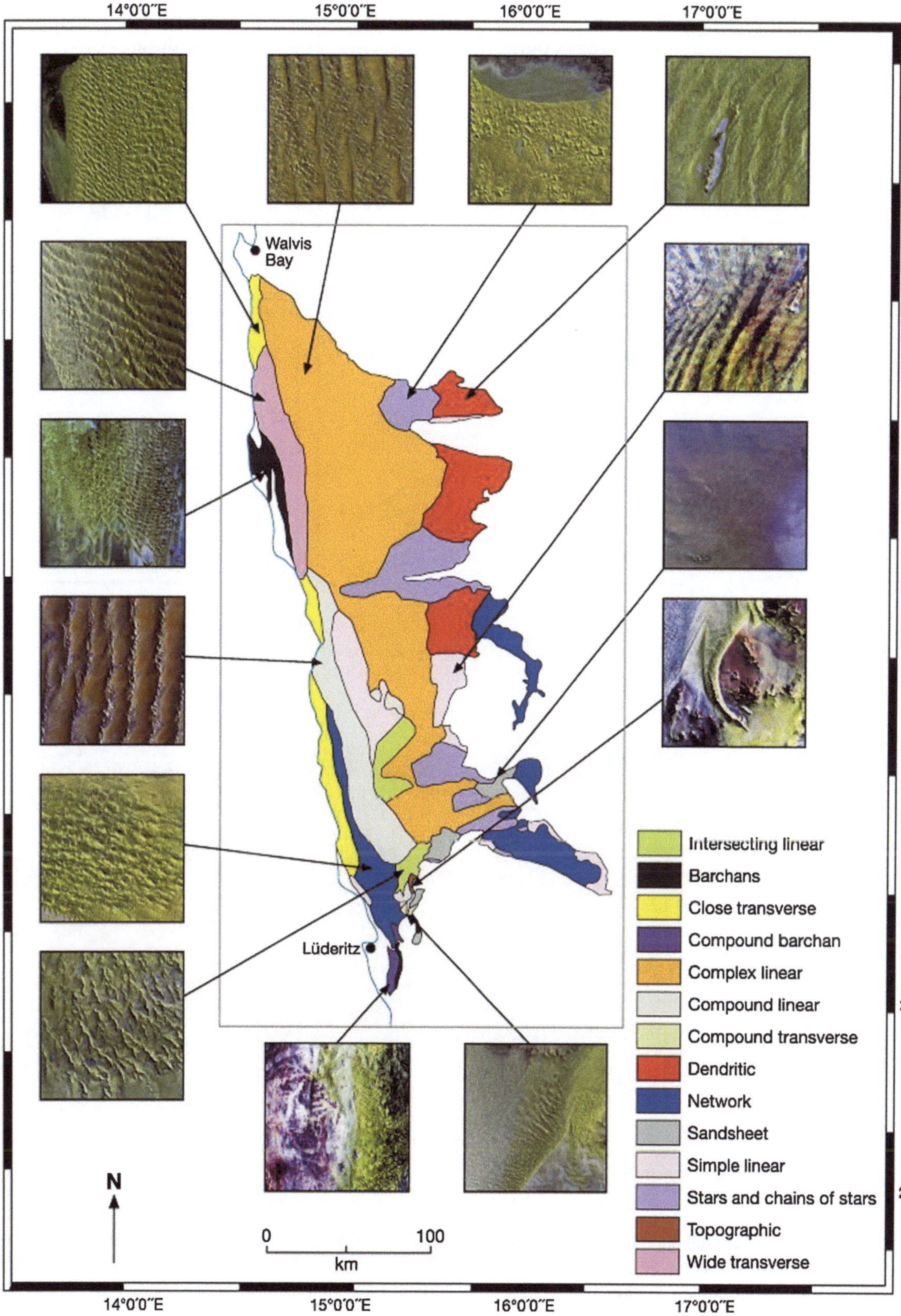

Fig. 18.2 Dune types in the Namib Sand Sea (from Livingstone et al. 2010, Fig. 2)

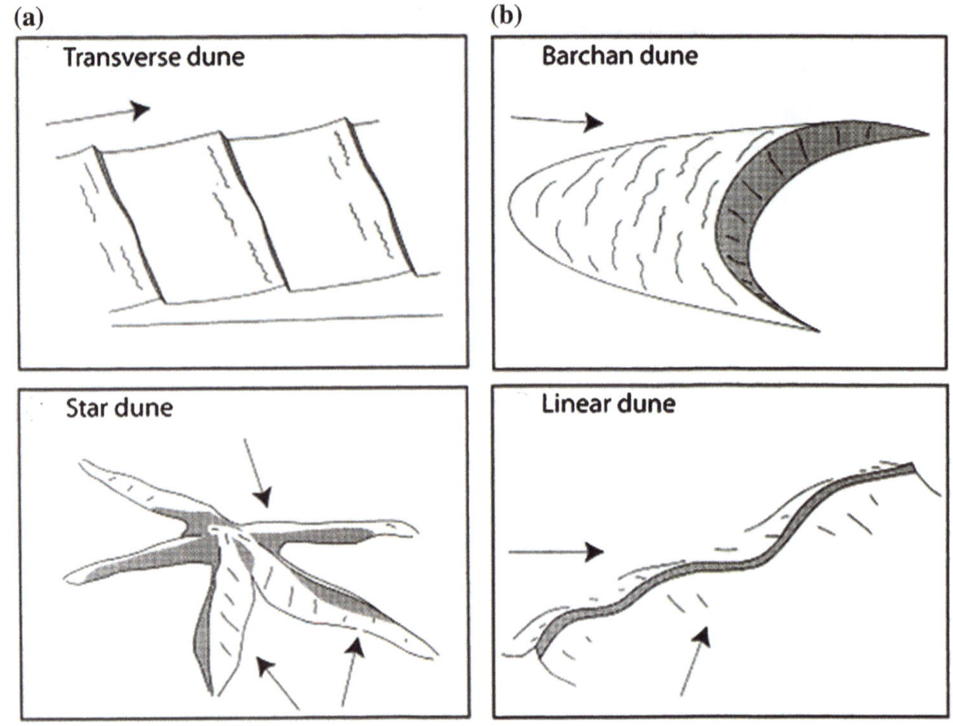

Fig. 18.3 Some major types of dune found in the Namib in relation to wind directions

Fig. 18.4 Linear dune near Gobabeb

the interior. The dominant annual sand movement appears to be from the south, as has been revealed by long-term monitoring of a small linear dune near Gobabeb (Besler et al. 2013).

The spacing of linear dunes varies through the sand sea. It is greatest in the central regions at 1,800–2,500 m, whereas in the southern parts these dunes are generally spaced at 1,500–2,000 m. The dunes are mostly between 6,000 and 900 m wide and between 50 and 150 m high. They are the dominant dune form in the sand sea, covering about 74 % of the sand sea area (Lancaster 1989). Changes in their form and the nature of wind flow over them during three decades have been described by Livingstone (1985, 2003). He found that the crests of the dunes move laterally back and forth in response to seasonal switching of wind direction, but return at the end of the year's cycle to their position at the beginning. This, says Livingstone, suggests that the dune is an equilibrium response to the wind regime and that there is no evidence of any lateral shifting of the dune. However, recent studies using Optically Stimulated Luminescence dating and Ground Penetrating Radar (Bristow et al. 2005, 2007) have suggested that a few hundreds of metres of lateral migration have taken place in the last few thousands of years.

Star dunes (Fig. 18.5), characterised by their pyramidal morphology and radiating sinuous arms, are associated with complex, multidirectional wind regimes (SW–WSW, NE–E and N) that occur along the sand sea's eastern margin, close to the Great Escarpment. The dunes are highest and most widely spaced in the central and some northern parts of the erg, with progressively lower and more closely spaced dunes towards the margins. Some of the star dunes, including those in proximity to the much visited Sossus Vlei, are well over 150 m high, and some may reach heights of 200–300 m, making them some of the largest dunes in the world, only exceeded perhaps by those of China's Badain Jaran Desert.

The sharply defined northern margin of the Sand Sea is formed by the Kuiseb River, for although this is only ephemeral, it flows sufficiently often and powerfully to prevent the dunes, driven by winds from the south, from moving further north, except in the immediate coastal fringe between Walvis Bay and Swakopmund. Further south, rivers deriving their flow from the mountains of the interior, such as the Tsauchab and the Tsondab Vleis, have entered the dune field and deposited light coloured, horizontally laminated silts, but at the present time they do not have the power to reach the Atlantic (Stone et al. 2010). If and when conditions were moister in the past, these rivers may have extended further into the sand sea than they do today, depositing lake sediments (Teller and Lancaster 1986).

The sources for the sand that make up the Sand Sea are probably varied and could include eroded material from the older Tsondab Sandstone Formation (Besler and Marker 1979), weathered debris from granites and Karoo sandstones carried down by streams from the interior (Besler 1984), deflation of sand from river beds, and derivation from the Atlantic shoreline. Lancaster and Ollier (1983) believe that much of the sand may have been supplied by the Orange River to the coastal zone and then been blown inland, and this has recently been confirmed by Garzanti et al. (2012) and Gehring et al. (2014).

18.4 The Source and Colour of the Dune Sands

The colour of the dune sand of the Namib shows clear spatial trends (Fig. 18.6). In coastal areas dominated by crescentic dunes the sand is yellowish brown to light yellowish brown, whereas in eastern areas it becomes a very striking yellowish red, a shade that would not disgrace a ripe apricot. Four main hypotheses have been suggested to account for the reddening of dunes as one progresses inland (Walden et al. 1996; Walden and White 1997). The first of these is that the increasing age of the sands inland

Fig. 18.5 Star dunes near Sossus Vlei

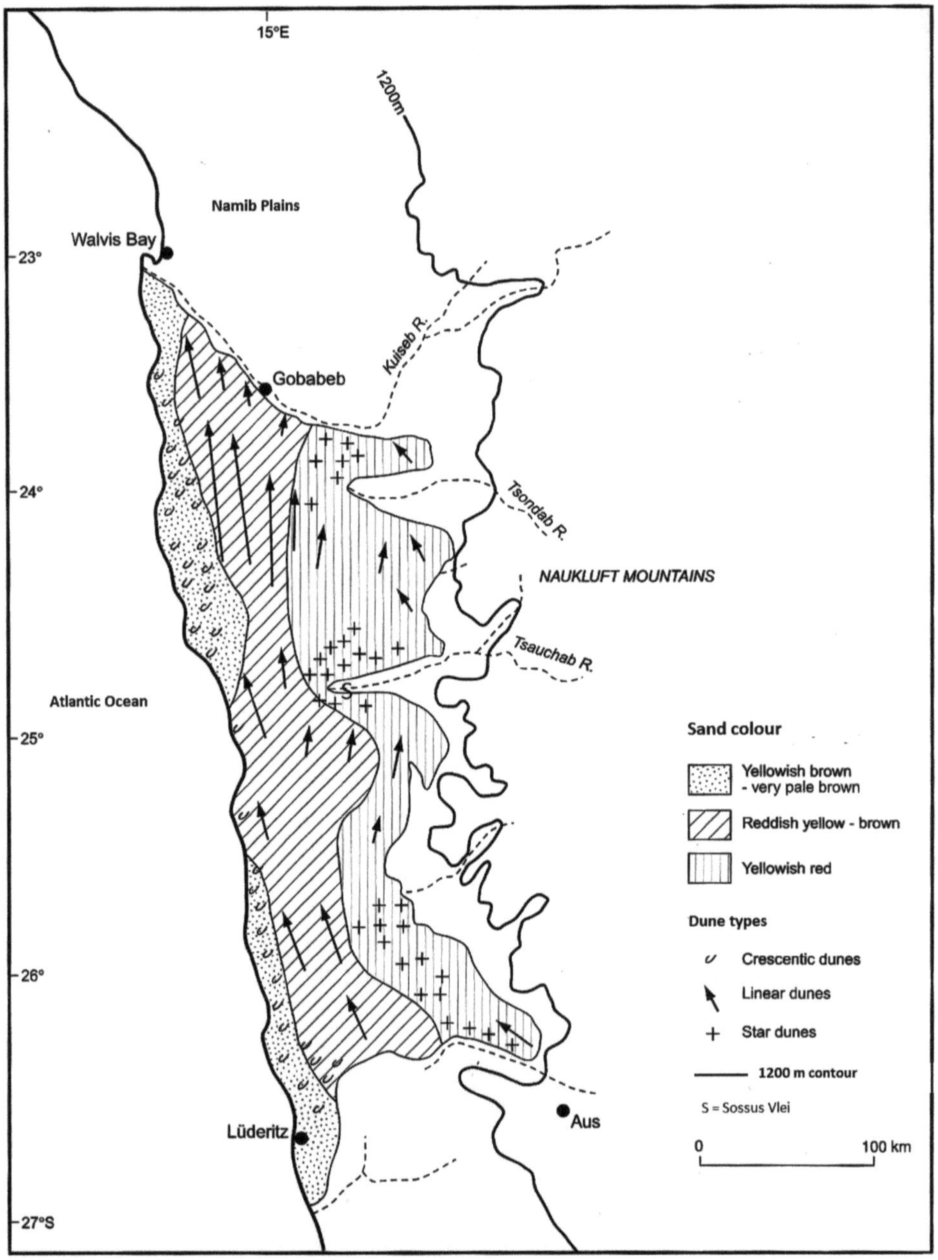

Fig. 18.6 Dune colour trends in the Namib Sand Sea (from Walden and White 1997, Fig. 1)

allows greater time for weathering processes to develop the iron (haematite) coatings around quartz grains. The second is that in areas of active sand transport and high energy winds near the coast, coatings may be lost or fail to develop. The third is that different sand source materials occur in the coastal zone compared with inland. Finally, it may be that a regional climatic gradient with warmer and wetter conditions inland provides a control on the rates of weathering processes which generate the haematite coatings. On the basis of detailed analyses, Walden and White (1997) suggest that different sand source materials play a major role, but that so also do age and environmental gradients.

References

Bagnold RA (1941) The physics of blown sand and desert dunes. Methuen, London

Barnes J (2001) Barchan dunes on the Kuiseb River delta, Namibia. S Afr Geogr J 83:283–292

Besler H, Marker ME (1979) Namib Sandstone: a distinct lithological unit. Trans Geol Soc S Afr 82:155–160

Besler H (1984) The development of the Namib dune field according to sedimentological and geomorphological evidence. In: Vogel JC (ed) Late Cainozoic plaeoclimates of the Southern Hemisphere. Balkema, Rotterdam, pp 445–453

Besler H, Lancaster N, Bristow C, Henschel J, Livingstone I, Seely M, White K (2013) Helga's dune: 40 years of dune dynamics in the Namib Desert. Geogr Ann A86:381–388

Bristow CS, Lancaster N, Duller GAT (2005) Combining ground penetrating radar surveys and optical dating to determine dune migration in Namibia. J Geol Soc 162:315–321
Bristow CS, Duller GAT, Lancaster N (2007) Age and dynamics of linear dunes in the Namib Desert. Geology 35:555–558
Bubenzer O, Bolten A (2008) The use of new elevation data (SRTM/ ASTER) for the detection and morphometric quantification of Pleistocene megadunes (draas) in the eastern Sahara and the southern Namib. Geomorphology 102:221–231
Bullard JE, White KW, Livingstone I (2011) Morphometric analysis of aeolian bedforms in the Namib Sand Sea using ASTER data. Earth Surf Proc Landf 36:1534–1549
Garzanti E, Ando S, Vezzoli G, Lustrino M, Boni M, Vermeesch P (2012) Petrology of the Namib Sand Sea: long-distance transport and compositional variability in the wind-displaced Orange Delta. Earth-Sci Rev 112:173–189
Gehring AU, Riahi N, Kind J, Almqvist BS, Weidler PG (2014) The formation of the Namib Sand Sea inferred from the spatial pattern of magnetic rock fragments. Earth Planet Sci Letters 395:168–172
Kocurek G, Lancaster N, Carr M, Frank A (1999) Tertiary Tsondab Sandstone: preliminary bedform reconstruction and comparison to modern Namib Sand Sea dunes. J Afr Earth Sci 29:629–642
Lancaster N (1989) The Namib Sand Sea: dune forms, processes and sediments. Balkema, Rotterdam
Lancaster N, Ollier CD (1983) Sources of sand for the Namib Sand Sea. Z Geomorph Suppl 45:71–83
Livingstone I (1985) The dynamics of sand transport on a Namib linear dune. Unpublished DPhil thesis, University of Oxford
Livingstone I (2003) A twenty-one-year record of surface change on a Namib linear dune. Earth Surf Proc Landf 28:1025–1031
Livingstone I, Bristow C, Bryant RG, Bullard J, White K, Wiggs GFS, Baas ACW, Bateman MD, Thomas DSG (2010) The Namib Sand Sea digital database of aeolian dunes and key forcing variables. Aeolian Res 2:93–104
Livingstone I (2012) Aeolian geomorphology of the Namib Sand Sea. J Arid Env 93:30–39
McKee ED (1982) Sedimentary structures in dunes of the Namib desert, south west Africa. Geol Soc Am Special Papers 188:1–2
Miller RMcG (2008) The geology of Namibia, vol 3. Geological Survey of Namibia, Windhoek
Morales J, Senut B, Pickford M (2011) *Crocuta dietrichi* from Meob, Namibia: implications for the age of the Tsondab Sandstone in the coastal part of the Namib Desert. Estudios Geol 67:207–215
Pickford M (2000) Fossil spider's webs from the Namib Desert and the antiquity of Seothyra (Araneae, Eresidae). Ann Paléontol 86:147–155
Rango A, Chopping M, Ritchie J, Havstad K, Kustas W, Schmugge T (2000) Morphological characteristics of shrub coppice dunes in desert grasslands of southern New Mexico derived from scanning LIDAR. Remote Sensing Environ 74:26–44
Segalen L, Renard M, Pickford M, Senut B, Cojan I, Le Callonnec L, Rognon P (2002) Environmental and climatic evolution of the Namib Desert since the Middle Miocene: the contribution of carbon isotope ratios in ratite eggshells. C R Geosci 334:917–924
Senut B, Dauphin Y, Pickford M (1998) New avian remains from the Neogene of the Sperrgebiet (Namibia): refinement of the avian biostratigraphy of Namib Desert aeolianites. C R Acad Sci Series IIA Earth Plan Sci 327:639–644
Slattery MJ (1990) Barchan migration on the Kuiseb River Delta, Namibia. S Afr Geogr J 72:5–10
Stone AEC (2013) Age and dynamics of the Namib Sand Sea: a review of chronological evidence and possible landscape development models. J Afr Earth Sci 82:70–87
Stone AEC, Thomas DSG, Viles HA (2010) Late Quaternary palaeohydrological changes in the northern Namib Sand Sea: new chronologies using OSL dating of interdigitated aeolian and water-lain interdune deposits. Palaeogeogr Palaeoclim Palaeoecol 288:35–53
Teller J, Lancaster N (1986) Lacustrine sediments at Narabeb in the central Namib desert, Namibia. Palaeogeogr Palaeoclim Palaeoecol 56:177–195
Vermeesch P, Fenton CR, Kober F, Wiggs GFS, Bristow CS, Xu S (2010) Sand residence times of one million years in the Namib Sand Sea from cosmogenic nuclides. Nature Geosci 3:855–862
Walden J, White K (1997) Investigation of the controls on dune colour in the Namib Sand Sea using mineral magnetic analyses. Earth Plan Sci Letters 152:187–201
Walden J, White K, Drake NA (1996) Controls on dune colour in the Namib Sand Sea: preliminary results. J Afr Earth Sci 22:349–353
Ward JD (1988) Eolian, fluvial and pan (playa) facies of the Tertiary Tsondab Sandstone Formation in the Central Namib Desert, Namibia. Sed Geol 55:143–162

Sossus Vlei and its Star Dunes

19

Abstract

Sossus Vlei occurs as the termination of the Tsauchab River. Its westward passage to the Atlantic has been blocked by northwards migrating dunes of the Namib Sand Sea. It is only in exceptionally wet years that the Tsauchab reaches Sossusvlei. When it does so, the dry silt pans of the valley (vlei) are replaced by shallow lakes. Star dunes rise up to 325 m above its floor. They appear to be preferentially developed in areas with multi-directional or complex wind regimes, and are often located close to topographic features, such as the Namibian Great Escarpment, which modify the regional wind regimes and tend to increase wind variability.

Sossus Vlei, first documented in 1909 by Lieutenant Walter Trenk of the German Schutztruppe, is one of the most famous and iconic sites in Namibia and lies in the Namib Sand Sea within the Namib Naukluft Park. It is the most visited landform site in the whole country. It is notable for its huge red dunes, the colour of which contrasts with the almost white lake or flood silts that lie in the Vlei itself, and the cobalt blue of the sky. Tourists can drive down the Tsauchab valley on a good road, admiring and photographing the dunes that occur on both sides, until they reach the Vlei, though some stop en route at about 45 km from Sesriem to climb Dune 45. At Dead Vlei, gaunt skeletons of camelthorn trees stand on the vlei silts, and most of them seem to have stated growing during a moist phase about 880 years ago. They may have survived for around 300 years before the climate deteriorated and killed them. The river derives its name from a Nama term for 'the river where there are many salsola bushes' and these plants, *Salsola aphylla*, known as ganna, are found all along the river and at Sossus Vlei itself. Some of the sandy areas on the drive to the vlei have fairy circles, the nature of which is described in Chap. 25.

Sossus Vlei is the current end point of the Tsauchab River, the westward passage of which to the Atlantic has been blocked by northwards migrating dunes of the Namib Sand Sea (Fig. 19.1). In the past its palaeochannel, which is locally marked by gravels to the west of Sossus Vlei, may have entered the ocean at Fischersbrunn, south of Meob Bay. The Tsauchab reached beyond its present end-point by 2–3 km at c 25,000 years ago, and again in the early Holocene at 9–7 thousand years ago (Brook et al. 2006). The Tsauchab today is a small, ephemeral stream which has its source to the east of the Naukluft Mountains. Before it reaches the sand sea it passes through a canyon that it has created, the 30 m deep Sesriem Canyon. This is cut into the lightly cemented, calcified gravels of the Karpenkliff Formation, which date back to c 15 million years ago (middle Miocene) and which were probably deposited on a former braided plain of a Proto-Tsauchab. It is only in exceptionally wet years that the Tsauchab reaches Sossusvlei so that the dry silt pans are replaced by shallow lakes. Recent examples of such years include 1963, 1974, 1976, 1987, 1997, 1999, 2000, 2006, 2008 and 2011. After a few months the vlei dries out and the pale, pan sediments crack into polygonal forms.

Star dunes rise up to 325 m above the floor of the vlei (Fig. 19.2). They rest in part on an elevated terrace cut into the Tsondab sandstone. They are some of the biggest dunes known on the face of Earth, though not the biggest as has sometimes been claimed. Star dunes (or *rhourds*) are dunes that have three or more arms extending radially from a central peak. Each arm has a steep-sided, sinuous crest, with avalanche faces. Typically between 1 and 2 km across, they are often the largest dune type found in many sand seas. They have been recorded from many deserts, but have until comparatively recently been the subject of relatively little research (Nielson and Kocurek 1987). They are particularly important in the Grand Erg Oriental of the Sahara (where

A. Goudie and H. Viles, *Landscapes and Landforms of Namibia*, World Geomorphological Landscapes,
DOI 10.1007/978-94-017-8020-9_19,

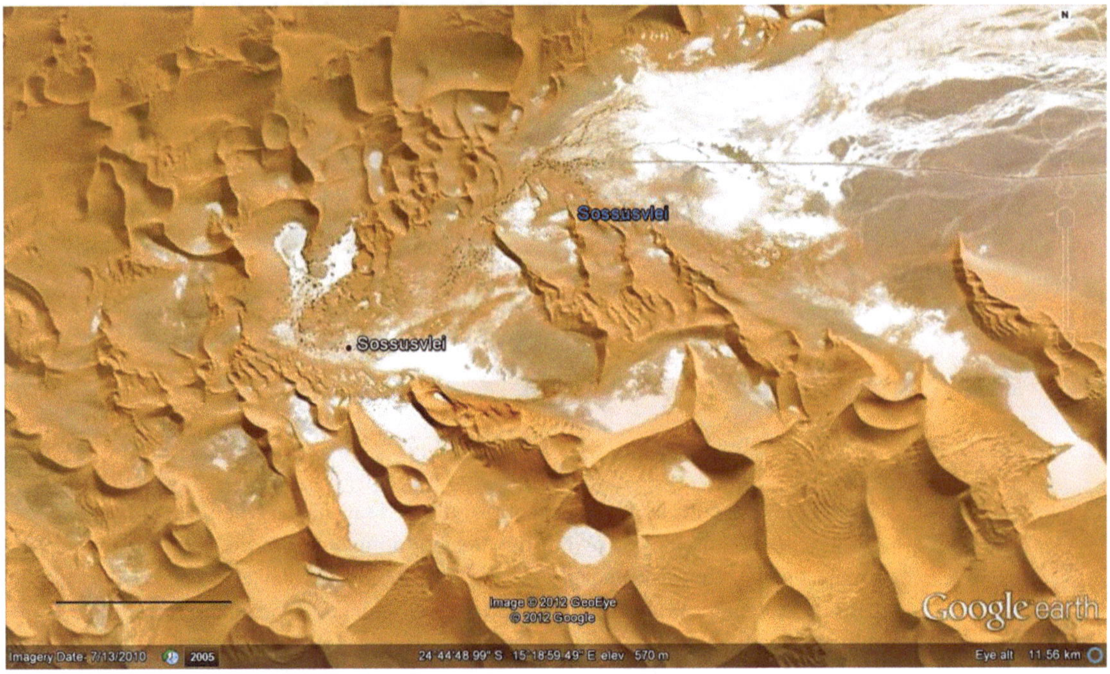

Fig. 19.1 Google Earth image of Sossus Vlei, silt pans and star dunes. Scale bar 2 km (© 2012 GeoEye, Google)

Fig. 19.2 Sossus Vlei, with nebkhas and white silts in the foreground

they comprise about 40 % of the dunes), on the east side of the Murzuk sand sea in Libya, the north end of the Erg Chech and the north end of the Erg Iguidi. Good examples also occur on the margins of the Rub Al' Khali in south Oman, and the Badan Jarain desert of China. However, they appear to be absent from the Australian, Kalahari and Thar Deserts and also from some of the Saharan ergs (Lancaster 1995, p. 71). They appear to be preferentially developed in areas with multi-directional or complex wind regimes, and are often located close to topographic features, such as the Namibian Great Escarpment, which modify the regional wind regimes and tend to increase wind variability.

References

Brook GA, Srivastava P, Marais E (2006) Characteristics and OSL minimum ages of relict fluvial deposits near Sossui Vlei, Tsauchab River, Namibia, and a regional climate record for the last 30 ka. J Quat Sci 21:347–362

Lancaster N (1995) Geomorphology of desert dunes. Routledge, London

Nielson J, Kocurek G (1987) Surface processes, deposits, and development of star dunes: Dumont dune field, California. Bull Geol Soc Am 99:177–186

The Naukluft Mountains and their Tufa Cascades

20

Abstract

The Naukluft Mountains occur on the western edge of the interior highlands and form part of the Great Escarpment. The dissolution of ancient carbonate rocks within these mountains, has caused streams draining the area to be heavily charged with dissolved calcium carbonate. As a result, large accumulations of freshwater calcium carbonate have been produced at points within the channel, often forming impressive cascades and barrages. These are known as tufa or travertine deposits, and their formation is often mediated and influenced by plants and microbes growing within the streams.

The Naukluft Mountains form one of the corner buttresses of the Great Escarpment (see Chap. 1) that marks the western edge of the interior highlands. They rise to as much as 1,965 m above sea level, and are composed of three main geological units: the metamorphic and intrusive rocks of the Rehoboth and Sinclair sequences (c 1,000–2,000 million years old), Nama group sediments (c 600 million years old) and the Naukluft Nappe Complex (Korn and Martin 1959), which was emplaced during the Damara Orogeny c 500–550 million years ago. Parts of this landscape are ancient, for the much-folded mountains are truncated by a great, inclined peneplain, so that the highest mountains are flat-topped, while the broad valleys, many of them U-shaped, may have been moulded by Dwyka ice. The Nama sediments, which were deposited in a shallow, tropical sea, include limestones and dolomites, and waters passing through these have first dissolved them, to produce caves (Irish et al. 2000), dolines and other karstic features, and then have caused deposition of tufa to occur (Figs. 20.1 and 20.2).

Tufa and travertine are terrestrial freshwater accumulations of calcium carbonate, whose formation often involves a degree of organic involvement. The names tufa and travertine can be used synonymously, but often tufa is taken to refer to a softer, more friable deposit whilst travertine refers to a harder, more resistant material frequently used as a building material. Tufas and travertines form in freshwater environments where thermodynamic and kinetic characteristics favour the precipitation of calcium carbonate from carbonate-rich waters. Such conditions arise where carbon dioxide is removed from the water through turbulent degassing, evaporation, or biological uptake. Suitable conditions for precipitation of tufa and travertine are often found in or near karst areas where dissolution of limestone provides high levels of dissolved carbonate, or where thermal waters, rich in carbon dioxide, originate in areas of recent volcanic activity. There has been much debate over the major controls on tufa and travertine formation and in particular the role of organisms in their genesis. Certainly, aquatic plants and microorganisms can aid deposition of tufa—by providing precipitation nuclei, by removing carbon dioxide from the water and perhaps also be direct precipitation of calcium carbonate—but in some environments physicochemical controls on precipitation outweigh any biological involvement. Within rivers tufas and travertines can form spectacular barrages often with waterfalls cascading over them, and with clastic tufa accumulating behind the barrage. In some fluvial environments, suites of lakes become created between barrages. Around springs, mounds and terraces can develop, and where springs debouch on steep slopes such deposits can form huge prograding cascades.

Tufa deposits are widespread in parts of the central Namib. They occur at various points along the Kuiseb Valley, where they are called the Hudaob Tufa Formation (Ward 1984) but have their best development in the Naukluft Mountains, where there are extensive cascade deposits (Figs. 20.3 and 20.4). Tufa is found commonly in both northward and southward draining catchments in the Naukluft—although the

A. Goudie and H. Viles, *Landscapes and Landforms of Namibia*, World Geomorphological Landscapes,
DOI 10.1007/978-94-017-8020-9_20, © Springer Science+Business Media Dordrecht 2015

Fig. 20.1 Blasskopf deposit, farm, Naukluft Moutains

Fig. 20.2 Actively depositing in the Naukluft Park

northward draining ones only experience relatively rare flows in today's climate, the southward draining ones have perennial pools and small amounts of flow over the barrages and cascades. The tufa deposits in the Naukluft Mountains show the imprint of highly variable flow regimes that cause differing facies associated with deposition, quiescence and erosion as explained by Viles et al. (2007) who suggest the following model of tufa formation to explain the different facies that occur:

1. An initial irregularity in the stream long profile creates turbulence in the stream which leads to carbonate deposition if water supply conditions are adequate. Moss also contributes to tufa deposition. A barrage gradually builds up. These barrages can be vast and complex features, with steeply angled cascade facies developed where water shoots over the top forming waterfalls. Pools develop behind the barrage so that laminated and reed facies develop in the quieter waters.

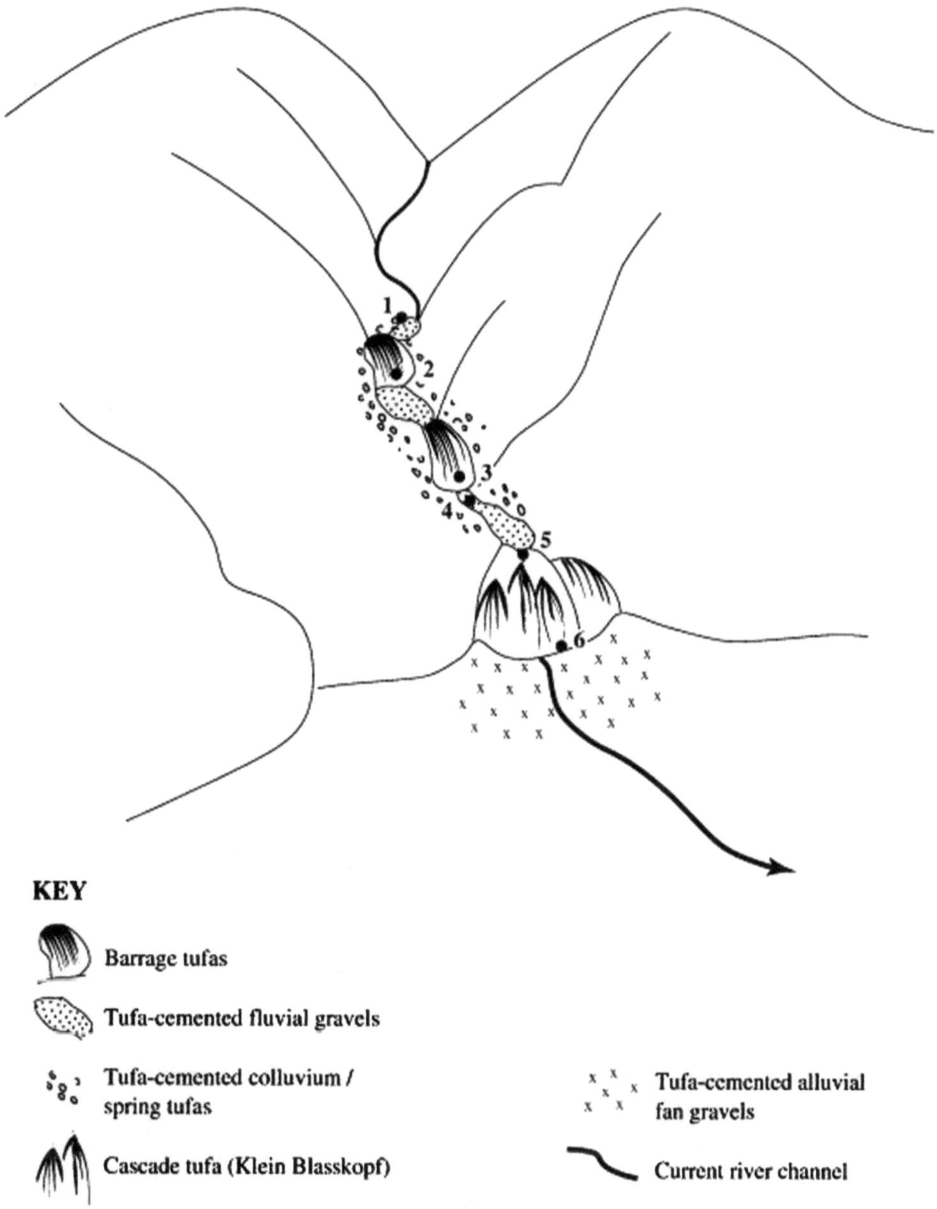

Fig. 20.3 Map of cascades upstream from Klein Blasskopf (from Viles et al. 2007, Fig. 2)

2. A phase of episodic high flows causes incision of the softer tufa barrages by large transported boulders. Total obliteration may sometimes occur, but normally remnants are preserved and the resultant deposits contain a mixture of boulders surrounded by tufas.
3. Subsequent lower flow conditions cause boulder deposition in the stream bed and tufa is deposited in their interstices.

As a result of repeated episodes of tufa build-up and erosion occurring over very long timespans quite complex barrage, cascade and channel tufa sequences can be developed (Fig. 20.4). At Blasskranz, for example, the Klein Blasskopf tufa cascade is some 80 m high and 400 m wide, whilst the main Blasskopf deposit is around 100 m high (Fig. 20.1). How old the deposits are is something of an enigma, but there is evidence for deposition within the Holocene and further back into the Pleistocene. Brook et al. (1999) present preliminary radiocarbon dates showing tufa deposition between 11,000 and 20,000 years ago, whilst Stone et al. (2010) report a reliable U/Th date of c 80,000 years ago for part of a large barrage in a southward draining catchment. Tufa is acknowledged to be complex to date using radiometric methods, because of multiple problems including detrital contamination and open systems behaviour, as Stone et al. (2010) acknowledge. Despite the relatively dry climate and low water flow rates, there is clear evidence of current active tufa deposition within southward draining streams within the Namib Naukluft Park (e.g. along the Waterkloof Trail section of the Naukluft River) associated with organic and inorganic influences.

Fig. 20.4 Model of cascade morphology (from Viles et al. 2007, Fig. 9)

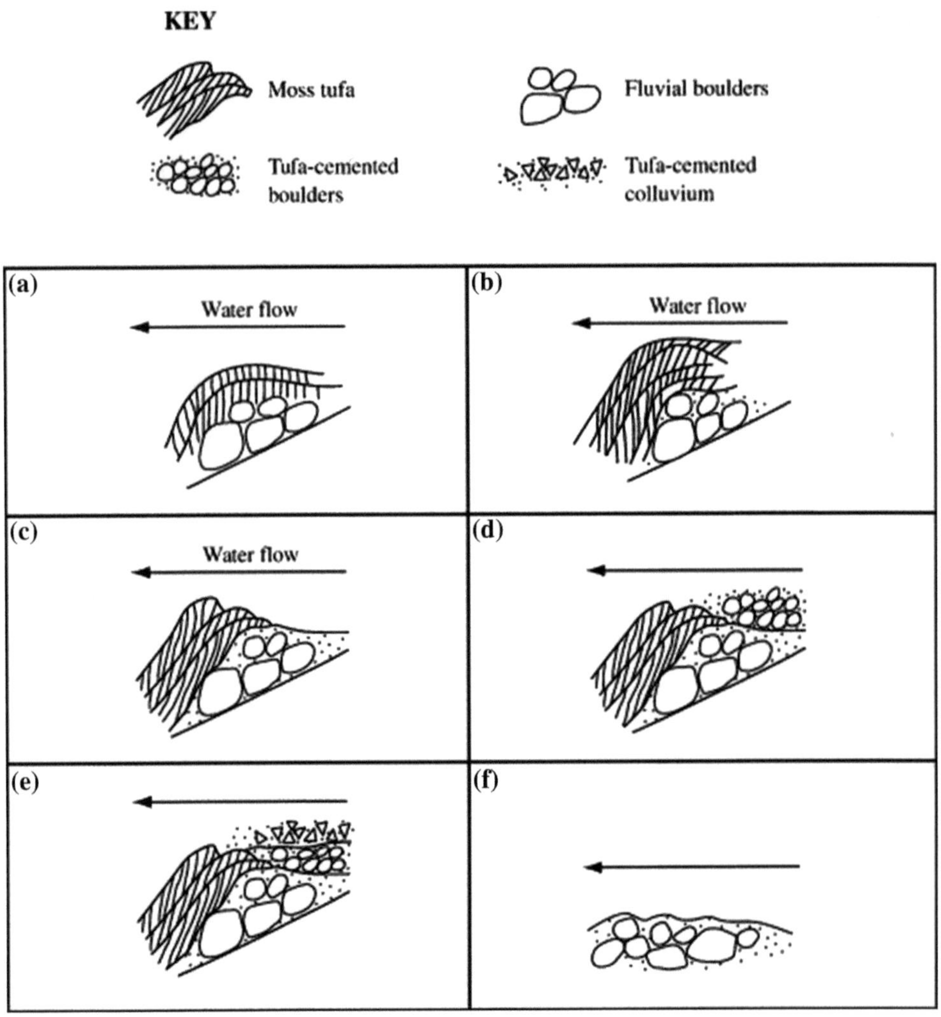

Elsewhere in central Namibia, the accumulation of salt may produce features akin to calcareous tufas. In the Swakop canyon, 8 km upstream from its confluence with the Khan, groundwater seepage has produced a 200 m long wall of tufa, festooned with stalactites and stalagmites. It encrusts plant material and ostrich feathers, and XRD analyses indicate that it is primarily composed of calcite and halite, with some silica.

References

Brook GA, Marais E, Cowart JB (1999) Evidence of wetter and drier conditions in Namibia from tufas and submerged speleothems. Cimbebasia 15:29–39

Irish J, Martini J, Marais E (2000) Cave investigations in Namibia VI. the desert caves in and around the Namib-Naukluft Park. Cimbebasia 16:177–193

Korn H, Martin H (1959) Gravity tectonics in the Naukluft Mountains of South West Africa. Geol Soc Am Bull 70:1047–1078

Stone AEC, Viles HA, Thomas L, van Calsteren P (2010) Can ^{234}U–^{230}Th dating be used to date large semi-arid tufas? challenges from a study in the Naukluft Mountains, Namibia. J Quat Sci 25:1360–1372

Viles HA, Taylor MP, Nicoll K, Neumann S (2007) Facies evidence of hydroclimatic regime shifts in tufa depositional sequences from the arid Naukluft Mountains, Namibia. Sed Geol 195:39–53

Ward JD (1984) A reappraisal of the Cenozoic stratigraphy in the Kuiseb Valley of the central Namib Desert. In: Vogel JC (ed) Late Cainozoic palaeoclimates of the Southern Hemisphere. Balkema, Rotterdam, pp 455–463

Linear Dunes of the Kalahari

21

Abstract

The Kalahari of eastern and northern Namibia has many linear sand dunes, some of which have tuning-fork junctions. In the north they run from east to west, while in the south west they generally run from northwest to southeast. Many of these dunes are currently inactive and may be the product of formerly drier conditions. In the southwest of the Kalahari they occur in an area where the rainfall is less than 250 mm per annum whereas in the northern areas the present rainfall well exceeds 1,000 mm. Those in the southwest have mobile crests in dry, windy years. In recent years a number of studies have been undertaken on the ages of the Kalahari dunes using thermoluminescence (TL) or optically stimulated luminescence (OSL) dating techniques. These have shown that there have been a number of phases of dune accumulation over at least the last 186,000 years and that the dunes of the south west Kalahari have been partially active during both the Late Pleistocene and the Holocene. The chapter discusses the various models that have been proposed to account for the formation of linear dunes.

21.1 Introduction

The dunes of the Kalahari are notable for a variety of reasons: they cover a vast area, most of them are inactive, few are very spectacular, most are linear dunes (Fig. 21.1), and many of those have particularly well developed 'tuning-fork junctions' (Fig. 21.2). They are composed of the famous, generally red and ocherous Kalahari Sand, which is dominantly quartzose and for the most part derived from local sources, including accumulated weathering products derived from Karoo and other rocks (Thomas and Shaw 1991, Sect. 3.4.1).

The dunes extend from the Upington region on the Orange River in South Africa far northwards into Angola, Zambia, Zimbabwe and the Congo (Fig. 21.3). They form a great anticlockwise wheelround not unlike the pattern of dunes found over Australia (Goudie 1970). Those in southeast Namibia trend more or less from northwest to southeast, though to the east of the Karas Mountains there is a small area where the dunes run from west to east. Those in northeastern Namibia run more or less from east to west. In the southwest of the Kalahari they occur in an area where the rainfall is less than 250 mm per annum whereas in the northern areas the present rainfall well exceeds 1,000 mm. Those in the southwest have mobile crests in dry, windy years, but over the rest of the Kalahari most of the dunes now appear to be stabilized by vegetation. The great bulk of the dunes, eighty five percent in the southwest (Fryberger and Goudie 1981), are linear types and this dominance is again reminiscent of Australia. The linear dunes, especially in the southwest, have 'tuning-fork' junctions that have a similar morphometry to dendritic stream systems (Goudie 1969) and which are seldom as well developed in other sand seas, except perhaps those of the Simpson Desert in Australia.

21.2 Origin of Linear Dunes

What are linear dunes and how do they form? The first point to make is that linear dunes, or *seifs*, are straightish ridges with slip faces on both sides that run more or less parallel to the resultant wind trend. Linear dunes are also sometimes called sand ridges or longitudinal dunes, but linear dune is now the preferred term, partly because it has no genetic connotations. They often develop a sharp crest which explains why they are called *seif* (a sword) in Arabic. They may also display a meandering tendency (Parteli et al. 2007).

A. Goudie and H. Viles, *Landscapes and Landforms of Namibia*, World Geomorphological Landscapes,
DOI 10.1007/978-94-017-8020-9_21, © Springer Science+Business Media Dordrecht 2015

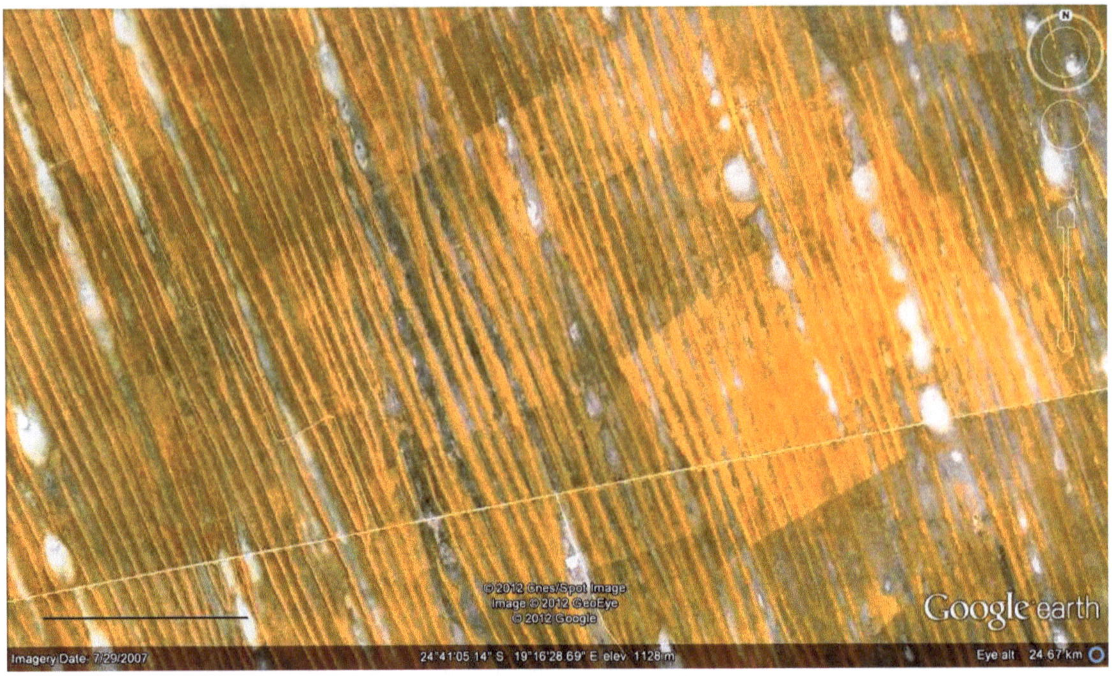

Fig. 21.1 Google Earth image of parallel linear dunes in the south west Kalahari. Scale bar 5 km (© 2012CNES/Spot Image, GeoEye, Google)

Fig. 21.2 Google Earth image of dunes with tuning-fork junctions in the south west Kalahari. Scale bar 1 km (© 2012 CNES/Spot Image, Google, AfriGIS (Pty) Ltd.)

They occur in loose sand in areas where there is seasonal or diurnal change in wind direction (Fig. 21.4)—i.e. a bimodal wind regime and where sand supply is relatively high (Parteli et al. 2009). They can also occur in areas with a more unimodal wind regime if the sand is locally stabilised by vegetation, sediment cohesion (due to the presence of salt, moisture or mud) or topographic shelter (Rubin and Hesp 2009).

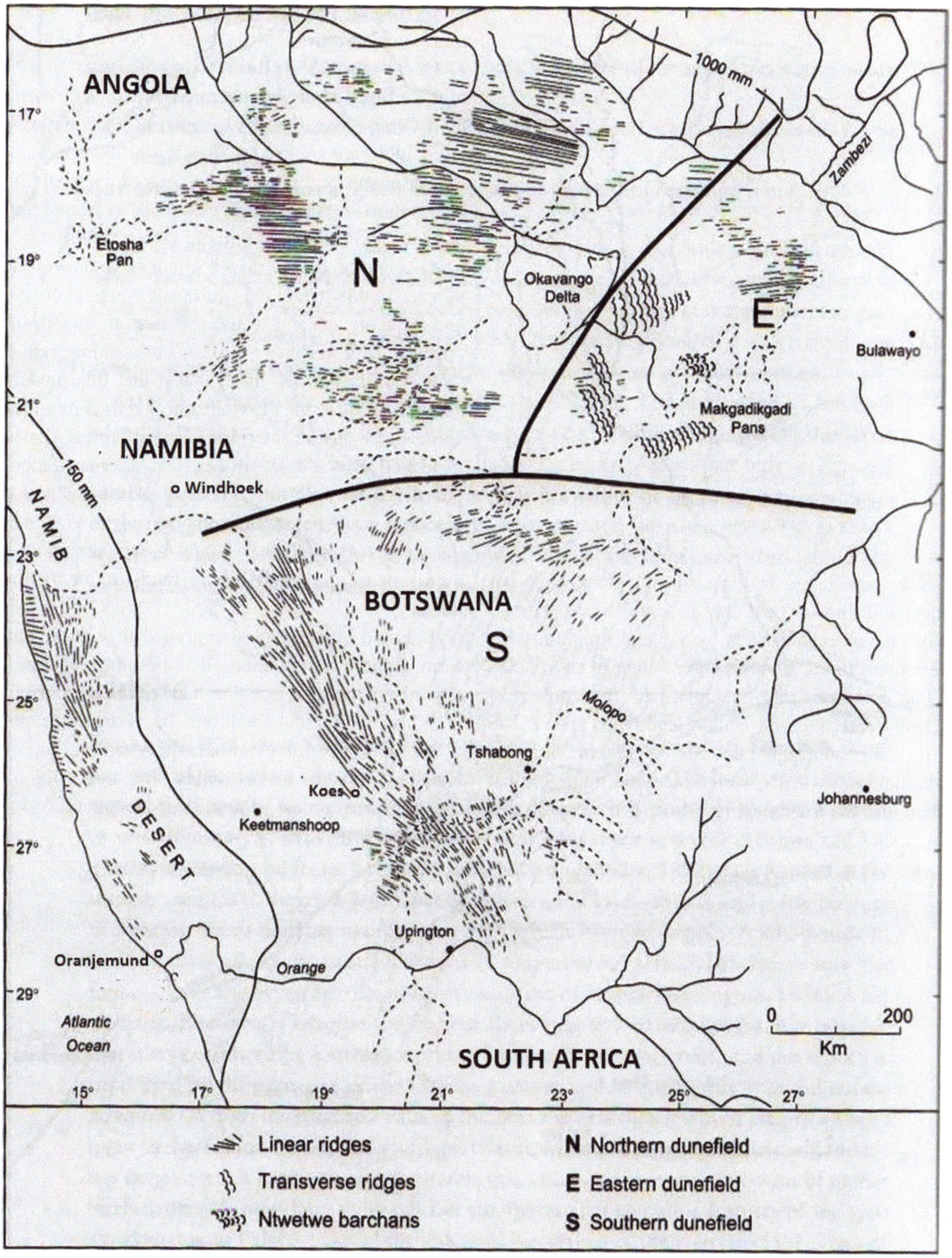

Fig. 21.3 The dunefields of the Mega-Kalahari. The 150 and 1,000 mm isohyets are shown

Some linear dunes may be modest in height (ten to twenty metres) and spacing (a few hundreds of metres) but others can be considerably larger, with heights in excess of 150 m, and a spacing of one or two kilometres. Examples of the former are the dunes of Australia and the Kalahari, whereas examples of the latter are the dunes of the central Namib (Chap. 18) or those of the Rub Al' Khali. The larger linear dunes are often described as complex or compound forms and may have multiple sub-dunes superimposed on a large plinth or draa. In general, as linear dunes get higher so they become more widely spaced (Lancaster 1995, p. 63).

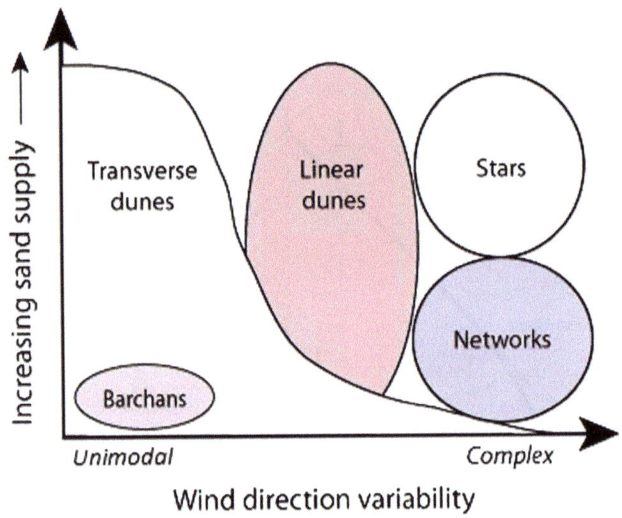

Fig. 21.4 The relationships between major dune types, availability of sand, and wind directional variability (after ideas of Wasson and Hyde, 1983)

There is a great range in the size and morphology of linear dunes, and this suggests that it would be an error to expect to be able to explain them by any one simple model. Some linear dune fields show a whole range of morphologies, and this is the case, for example, with the south west Kalahari (Bullard et al. 1995; Bullard and Nash 1998). The presence of river channels may modify the patterns of linear dunes as is the case with the valley-marginal dunes in that area (Bullard and Nash 2000) (Fig. 21.5). They may do this by modifying wind flow and by providing a local source of sand supply from their dry beds.

One early theory for linear dunes was that they were moulded by *thermally-generated helical roll vortices* (Bagnold 1953). These vortices, sometimes known as Langmuir circulation, are created by shearing in the boundary layer of the atmosphere. Bagnold suggested, speculatively, that paired, horizontal roll vortices, whose axes are parallel to the dominant wind direction, might sweep sand out of interdune troughs and onto sand ridges where currents would meet and ascend. In this model the wind pattern would create the dune and the dune spacing would represent the width of a pair of vortices. Roll vortices do exist, but a number of arguments have been developed that suggest that this model is not of general applicability (Livingstone 1988). First, there is little coincidence in the lateral spacing of linear dunes and the measured sizes of roll vortices. The latter are generally much greater. Secondly, roll vortices display measured transverse velocities well below that required to move sand. Thirdly, the model requires that winds blow parallel to the dune trend, an event which occurs rarely in many linear dune fields.

Bagnold (1941) had also argued that linear dunes could form from barchan dunes that became deformed as they moved into a regime which had less unimodal winds. This may happen in local situations, but scarcely seems a model that can apply, for example, in Australia, or, indeed the Kalahari, where linear dunes are near ubiquitous but barchans are almost absent. Verstappen (1968) suggested

Fig. 21.5 Google Earth image of river bordering dunes in the Kalahari. Scale bar is 1 km. The river valley runs on the north side of the road (© 2012CNES/Spot Image, Google)

that in the Thar Desert of India linear dunes could arise from the progressive elongation of parabolic dunes. This model may again have local applicability, but many linear dunes occur in areas where parabolic dunes are absent, which is the case in most of the Namib.

Modern models, based on field measurement of wind directions and velocities, relate linear dune development to the effects of bimodal wind regimes. Notable here are the studies of Livingstone in the Namib Sand Sea (1989, 1993). He showed that the crest of the linear dunes migrated laterally in response to seasonally bimodal wind regimes, but that net sand transport was along the dunes.

Assuming that linear dunes result from the operation of bimodal wind regimes, there are two different ways in which they may develop. On the one hand there is the downwind extension model which envisages that linear dunes extend longitudinally by progradation along their length. They are fed by sediment from upwind and the dunes become progressively younger downwind. Telfer (2011) found evidence for this in the southwest Kalahari, by dating a linear dune extending into a pan.

21.3 The Age of Kalahari Linear Dunes

In recent years a number of studies have been undertaken on the ages of the Kalahari dunes using thermoluminescence or optically stimulated luminescence dating techniques. They have demonstrated that there have been a number of phases of dune accumulation (Thomas et al. 2000; Stone and Thomas 2008) over at least the last 186,000 years and that the dunes of the south west Kalahari have been partially active during both the Late Pleistocene and the Holocene (Blümel et al. 1998).

In future decades, with global warming, it has been postulated that many of the linear dunes of the Mega-Kalahari will become much more active in response to a reduced vegetation cover resulting from lower precipitation and greater moisture loss through evapotranspiration (Thomas et al. 2005).

References

Bagnold RA (1941) The physics of blown sand and desert dunes. Methuen, London

Bagnold RA (1953) The surface movement of blown sand in relation to meteorology. Res Counc Israel, Spec Pub 2:89–93

Blümel WD, Eitel B, Lang A (1998) Dunes in southeastern Namibia: evidence from Holocene environmental changes in the southwestern Kalahari based on thermoluminescence data. Palaeogeogr Palaeoclim Palaeoecol 138:139–149

Bullard JE, Nash DJ (1998) Linear dune pattern variability if the vicinity of dry valleys in the southwest Kalahari. Geomorphology 23:35–54

Bullard JE, Nash DJ (2000) Valley-marginal sand dunes in the southwest Kalahari: their nature, classification and possible origins. J Arid Env 45:369–383

Bullard JE, Thomas DSG, Livingstone I, Wiggs GFS (1995) Analysis of linear sand dune morphological variability, southwestern Kalahari Desert. Geomorphology 11:189–203

Fryberger SG, Goudie AS (1981) Arid geomorphology. Progr Phys Geogr 5:420–428

Goudie AS (1969) Statistical laws and dune ridges in southern Africa. Geogr J 135:404–406

Goudie AS (1970) Notes on some major dune types in southern Africa. S Afr Geogr J 52:93–101

Lancaster N (1995) Geomorphology of desert dunes. Routledge, London

Livingstone I (1988) New models for the formation of linear sand dunes. Geography 73:105–115

Livingstone I (1989) Monitoring surface change on a Namib linear dune. Earth Surf Proc Land 14:317–332

Livingstone I (1993) A decade of surface change on a Namib linear dune. Earth Surf Proc Land 18:661–664

Parteli EJR, Durán O, Herrmann HJ (2007) Minimal size of a barchan dune. Phys Rev E 75:011301

Parteli EJR, Durán O, Tsoar H, Schwämmle V, Herrmann HJ (2009) Dune formation under bimodal winds. Proc Nat Acad Sci 106:22085–22089

Rubin DM, Hesp PA (2009) Multiple origins of linear dunes on Earth and Titan. Nat Geosci 2:653–658

Stone AEC, Thomas DSG (2008) Linear dune accumulation chronologies from the southwest Kalahari, Namibia: challenges of reconstructing late Quaternary palaeoenvironments from aeolian landforms. Quat Sci Rev 27:1667–1681

Telfer MW (2011) Growth by extension, and reworking, of a southwestern Kalahari linear dune. Earth Surf Proc Land 30:1125. doi:10.1002/esp.2140

Thomas DSG, Shaw PA (1991) The Kalahari Environment. Cambridge University Press, Cambridge

Thomas DSG, O'Connor PW, Bateman MD, Shaw PA, Stokes S, Nash DJ (2000) Dune activity as a record of late Quaternary aridity in the northern Kalahari: new evidence from northern Namibia interpreted in the context of regional arid and humid chronologies. Palaeogeogr Palaeoclimatol Palaeoecol 156:243–259

Thomas DSG, Knight M, Wiggs GFS (2005) Remobilization of southern African desert dune systems by twenty-first century global warming. Nature 435:1218–1221

Verstappen HTh (1968) On the origin of longitudinal (seif) dunes. Z Geomorph 12:200–220

Wasson RJ, Hyde R (1981) Factors determining desert dune type. Nature 304:337–339

The Weissrand and the Aligned Drainage of the Western Kalahari

22

Abstract

In southeastern Namibia is a large area of aligned terrain consisting of broadly parallel drainage lines, running approximately from NW to SE, and along which there are a multitude of small closed depressions, called dayas. The depressions have developed on the extensive calcrete deposits of the region. The spacing and alignment of the drainage lines implies (i) that they may have been superimposed onto the calcrete by a formerly more extensive dune cover and (ii) that the closed depressions may have developed along the drainage lines because of localized solution of the underlying calcrete, assisted perhaps by other pan-forming processes such as deflation and animal activity.

Analysis of Landsat 7 images of southeastern Namibia shows that there is a large area of aligned terrain (Fig. 22.1). It consists of broadly parallel drainage lines, running approximately from NW to SE, and along which there are a multitude of small closed depressions, called dayas (Goudie 2010), identification of which on the ground is far from easy, given the great subtlety of the relief. There is also an area of small dayas in the Ubib embayment of the Central Namib (Marker 1982), and similar features exist in the Aminuis area in the Kalahari.

The depressions have developed on the extensive calcrete deposits of the region, which cap Ecca (Permian) sedimentary rocks (which include shales and mudstones) and Kalahari Bed sediments, which include calcrete. The pedogenic calcrete profiles of the southwestern Kalahari can attain considerable thicknesses—sometimes over 30 m. The age of the Weissrand calcretes is probably early Tertiary and they are associated with the development of the African Surface (Mabbutt 1955; Blümel and Eitel 1994). Some of the calcrete contains fluvial pebbles, which form a scatter over the surface where they have been liberated from the calcrete congolomerate by weathering. Although cal-silcretes and silcalcretes occur, most deposits are dominated by calcium carbonate, with average calcium carbonate contents comprising ~70–80 % (Goudie 1973). Thick, pure calcretes, like other limestones, may therefore be susceptible to karstification which involves the dissolution of calcium carbonate to produce distinctive erosional terrain, characterized by sinkholes.

The spacing and alignment of the drainage lines on the Plateau implies (i) that they may have been superimposed onto the calcrete by a formerly more extensive dune cover and (ii) that the closed depressions may have developed along the drainage lines because of localized solution of the underlying calcrete, assisted perhaps by other pan-forming processes such as deflation and animal activity.

In southeastern Namibia, the best developed aligned drainage occurs in a zone to the east of Marienthal and Keetmanshoop and to the west of the Auob River, at ~26°S and 19°E. This area is known as the Weissrand. It covers a large tract of country that runs approximately 150 km from northwest to southeast and 50 km from southwest to north east. On its western side, particularly in the north, it is bounded by a clear escarpment, though further to the south this becomes much less obvious. The Weissrand is a plateau with very limited slopes, and mostly lies at an altitude of around 1,000–1,200 m above sea level. To the west it is bounded by a depression which drains to the Fish River and its tributaries, one of which, the Löwen, also forms the southern boundary to the plateau. To the north, near Marienthal, there is a triangular outlier to the Plateau, separated from the main body by the drainage of Die Vlak. To the east lie the main linear dunes of the southwest Kalahari, most of which are ~10–20 m in height. The aeolian dunes are dominantly linear forms, many with tuning fork junctions (see Chap. 21). Their alignment is predominantly from northwest to southeast, and this is exactly the same

A. Goudie and H. Viles, *Landscapes and Landforms of Namibia*, World Geomorphological Landscapes,
DOI 10.1007/978-94-017-8020-9_22, © Springer Science+Business Media Dordrecht 2015

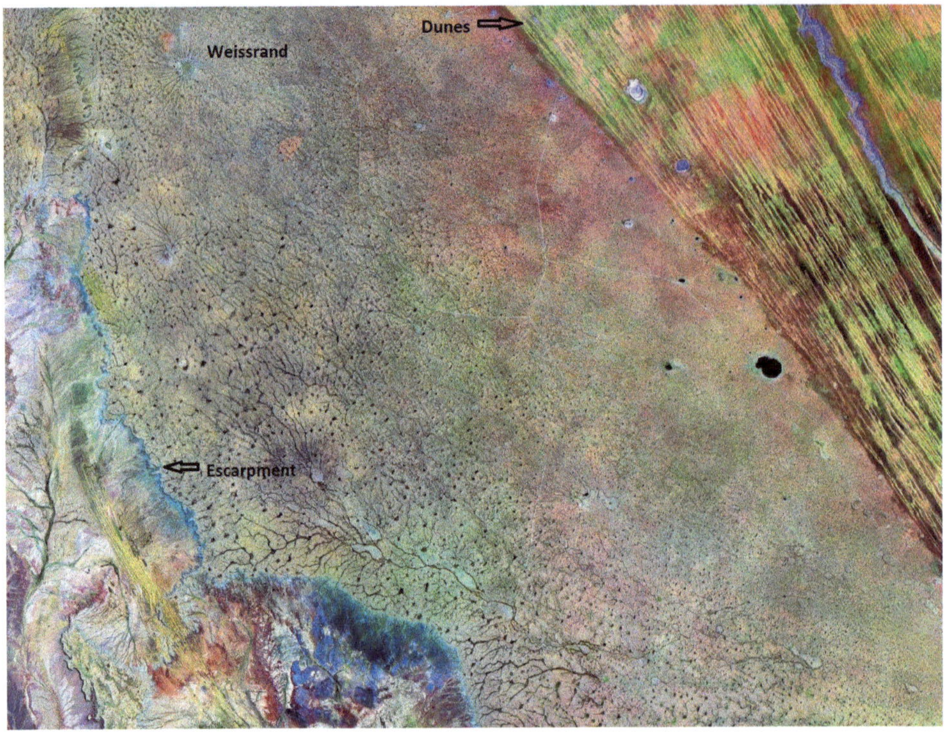

Fig. 22.1 Landsat 7 image of the Weissrand, showing linear dunes in the east, pans, and aligned dayas (courtesy of NASA)

alignment as the drainage developed on the calcrete surface of the Plateau. The dunes may have been more active in the Late Quaternary.

The depressions occur in large numbers. There are approximately 2 per square kilometer. They tend to occur in lines running from northwest to southeast, with around 1.3–1.4 km between depressions. The alignments in which they occur are spaced at approximately 1 km intervals. The depressions themselves have long axis lengths that range between about 100 and 1,200 m, with the mean being around 460 m.

There is a striking near correspondence between the alignment of the Weissrand drainage and the alignment of the linear dunes and inter-dunal depressions of the Kalahari. Both run from northwest to southeast. It is this near correspondence which suggests that the aligned drainage alignment has been imprinted from a former more extensive cover of Kalahari dunes. This cover formerly extended further west than it does today, and has now largely been stripped off, though small patches remain. Patches of linear dunes still occur on the downwind sides of depressions like that at Koes, and at the nearby salt pan at Vertwall.

The nature of the depressions (dayas) on the Weissrand Plateau (Fig. 22.2) and their spatial arrangement, suggests that they may have evolved through the following four stages (Goudie 2010):

(1) In Stage 1 there is an extensive calcrete surface over which there is a complete cover of branching linear dunes. Some very shallow depressions develop along the interdune swales either as a result of deflation or solution, or a combination of the two, but most of the calcrete is not directly exposed to the atmosphere.
(2) In Stage 2 the dune cover becomes stripped from the western part of the plateau, either because of a reduction in sand supply, or because of a shift in wind direction. This exposes the calcrete surface to the atmosphere. The depressions formed in the Stage 1 form an aligned pattern etched into the calcrete surface.
(3) In Stage 3 the depressions are developed further as a result of karstic processes, and themselves start to become foci for shallow drainage lines, which means that a more dendritic pattern develops. Such a dendritic pattern is accentuated near the escarpment edge, when drainage cuts back into the escarpment from the Fish River System, which lies to the west.
(4) In Stage 4 when the calcrete caprock becomes pierced, the underlying Ecca shales (Prince Albert Formation) are subjected to pan development by such processes as salt weathering and deflation, and eventually large, characteristically shaped pans evolve, some with lunettes on their lee sides, as shown by Koes Pan (see Chap. 23). A centripetal drainage pattern develops, focused on the pan.

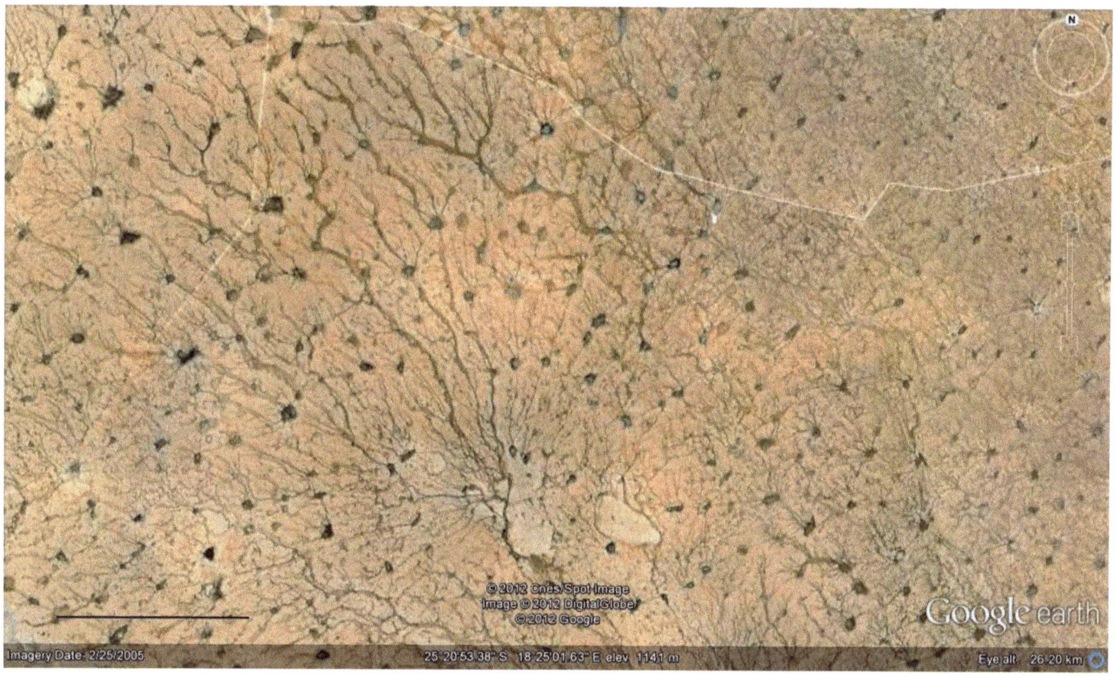

Fig. 22.2 Google Earth image of Weissrand dayas. Scale bar 5 km (© 2012 CNES/Spot Image, Digital Globe, Google)

Fig. 22.3 The Finger of God (Mukorob) before its collapse in 1988

One very distinctive landform that used to exist on the western side of the Weissrand was a very slim and precarious rock pillar called Mukorob (the Finger of God) (Fig. 22.3). The sandstones and mudstones that the Mukorob was developed in belong to the Permian Prince Albert Formation of the Karoo Sequence (270 million years old). The structure was 12 m high and up to 4.5 m wide, and weighed some 450 tons. Its base, developed in erodible mudstone, just 3 m long and 1.5 m wide, was narrower than the mass of rock which it supported. With the soft mudstone neck eaten away at a greater rate than the sandstone head, the neck eventually became too thin to support the head. It collapsed in December, 1988, perhaps because of an earth tremor or because of a rainstorm, thus depriving Namibia of one of its most visited landforms.

References

Blümel WD, Eitel B (1994) Tertiäre Deckschichten und Kalkrusten in Namibia: Entstehung und geomorphologische Bedeutung. Z Geomorp 38:385–403

Goudie AS (1973) Duricrusts in tropical and subtropical landscapes. Oxford University Press, Oxford

Goudie AS (2010) Dayas: distribution and morphology of dryland solutional depressions developed in limestones. Z Geomorph 54:145–159

Mabbutt JA (1955) Erosion surfaces in Namaqualand and the ages of surface deposits in the south-western Kalahari. Trans Geol Soc S Afr 58:13–29

Marker M (1982) Aspects of Namib geomorphology: a doline karst. Palaeoecol Afr 15:187–199

Koes and the Nature and Origin of Pans

23

Abstract

In the western Kalahari there is a group of classic pans. Such closed depressions are widespread in the Kalahari. Their origin has generated a large literature and hypotheses for their formation have included solution, excavation by animals, karstic and pseudo-karstic solution, and tectonic subsidence. However, that most pans are at least in part of aeolian origin is indicated by their distinctive morphology, their orientation with regard to prevailing wind directions, the bulbous shore on their lee sides, the presence of lunettes (composed in part of sediment deflated from pan floors) on their lee sides and observations on the ground and from space of dust plumes blowing from their surfaces. Lunette dunes are composed of mixtures of clay, silt, sand and salts derived from the pan floor, and there may be several generations evident.

In the western Kalahari, on the Weissrand Plateau, there is a group of classic pans of which Koes is the most impressive and accessible (Fig. 23.1). It has well developed lunette dunes forming mounds on its southeast margin, and a cluster of linear dunes running away from it in a south easterly direction. It is about 4.3 km across (from SW–NE), and has a classic bulbous shore on its south east side. Its lowest point, at about 964 m above sea level is c 50–60 m lower than the plateau into which is it cut. Heading south eastwards from Koes are a number of other pans, all of which have a similar morphology and all of which have a cordon of lunettes and linear dunes on their south east sides.

Such closed depressions are widespread in Namibia (Fig. 23.2), especially in the Kalahari. Large numbers, oriented in response to north westerly winds, occur in an immense tract of country to the east of Karasburg, Keetmanshoop and Marienthal towards the Botswana border. Another rather smaller group occurs in the vicinity of Tsumkwe in north east Namibia, and yet another large tract of pans occurs in the Ohangwena and Oshitoko districts between Etosha Pan and the Angolan border. These appear to have become oriented in response to winds blowing from the east.

The origin of pans has generated a large literature and hypotheses for their formation have included solution, excavation by animals (elephants, hogs, etc.), karstic and pseudo-karstic solution, and tectonic subsidence (Goudie and Wells 1995). That pans are at least in part of aeolian origin is indicated by their distinctive morphology (it has often been likened to a pork chop) with a crenulated windward side and a bulbous concave lee side, their orientation with regard to prevailing wind directions, the presence of lunettes (composed in part of sediment deflated from pan floors) on their lee sides (as is the case at Koes and its neighbours) and observations on the ground and from space of dust plumes blowing from their surfaces (Vickery et al. 2013). That said, other processes may contribute to the development of closed depressions in desert areas, including solution and animal excavation. The last process may have been more potent before huge herds of wild animals were decimated by human hunting (Alison 1899).

Paul Shaw and David Thomas, two desert geomorphologists with extensive Kalahari experience, have suggested that desert depressions originate through four main types of process: structural controls (e.g. faulting and rifting, and

A. Goudie and H. Viles, *Landscapes and Landforms of Namibia*, World Geomorphological Landscapes,
DOI 10.1007/978-94-017-8020-9_23,

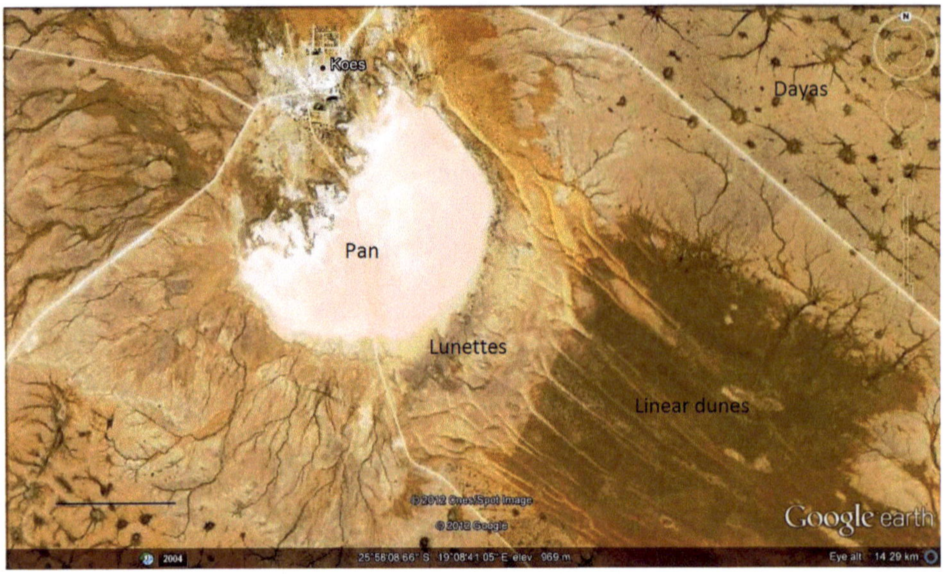

Fig. 23.1 Google Earth Image of Koes Pan. Scale bar 2 km (© 2012CNES/Spot Image, Google)

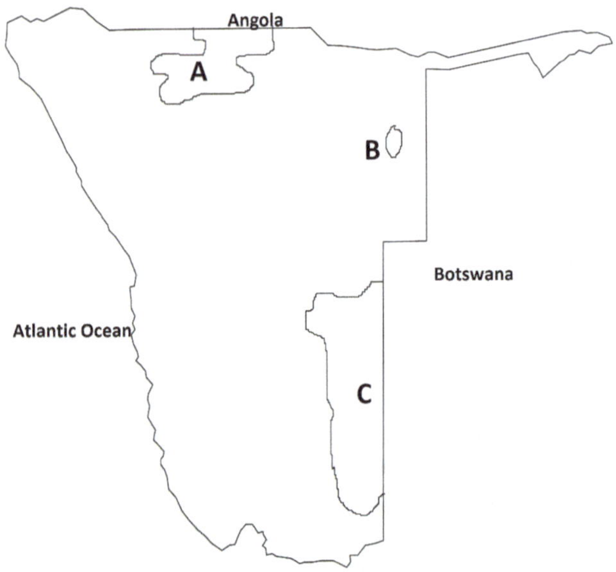

Fig. 23.2 The three main areas of inland pans in Namibia. *A* northern area, *B* Tsumkwe area, *C* south western Kalahari

downwarping); erosional controls (e.g. deflation, solution, animal scouring); ponding (e.g. in interdune troughs or ephemeral rivers) and dramatic (e.g. meteorite impacts, volcanic cratering) (Shaw and Thomas 1997). Pans are a feature of low angle surfaces on which the development of integrated surface drainage is limited. Whilst they probably result from a variety of processes, aeolian excavation is probably the most important. Lunette dunes, composed of excavated material, frequently (though not invariably) occur on their lee sides. In addition to a rainfall control on their distribution (they are a feature of semi-arid areas) there is also a strong control exercised by surface materials. The pans of southern Africa tend to occur on Ecca Shales, which weathers into fine-grained material, and Kalahari Sands, which are essentially friable. Both are therefore deflated relatively easily compared to stronger or coarser materials.

Goudie (1999) developed a model of pan development (Fig. 23.3) which both recognized the variety of formative influences, and classified them into various categories. First of all, there is the *predisposing condition* of low precipitation which has various consequences: vegetation cover is limited and so deflation can occur; animals concentrate at pans, causing trampling and overgrazing which also promote deflation; and salt accumulation occurs so that salt weathering can attack the fine-grained bedrock in which the depression lies, producing rock flour which can then be evacuated by the wind (Goudie and Viles 1997). These are *accentuating processes*, which serve to enlarge hollows, whether they are formed by other *initial formative processes* such as solution of carbonate and gypsum beds, or tectonics. It is also important if pans are to develop that the initial surface depression is not obliterated by the action of integrated or effective fluvial systems. *Non-restraining conditions* that limit fluvial activity are low angle slopes, episodic desiccation and dune encroachment, the presence of dolerite intrusions and tectonic disturbance. In addition it is important that pans do not lie in areas of active sand accumulation

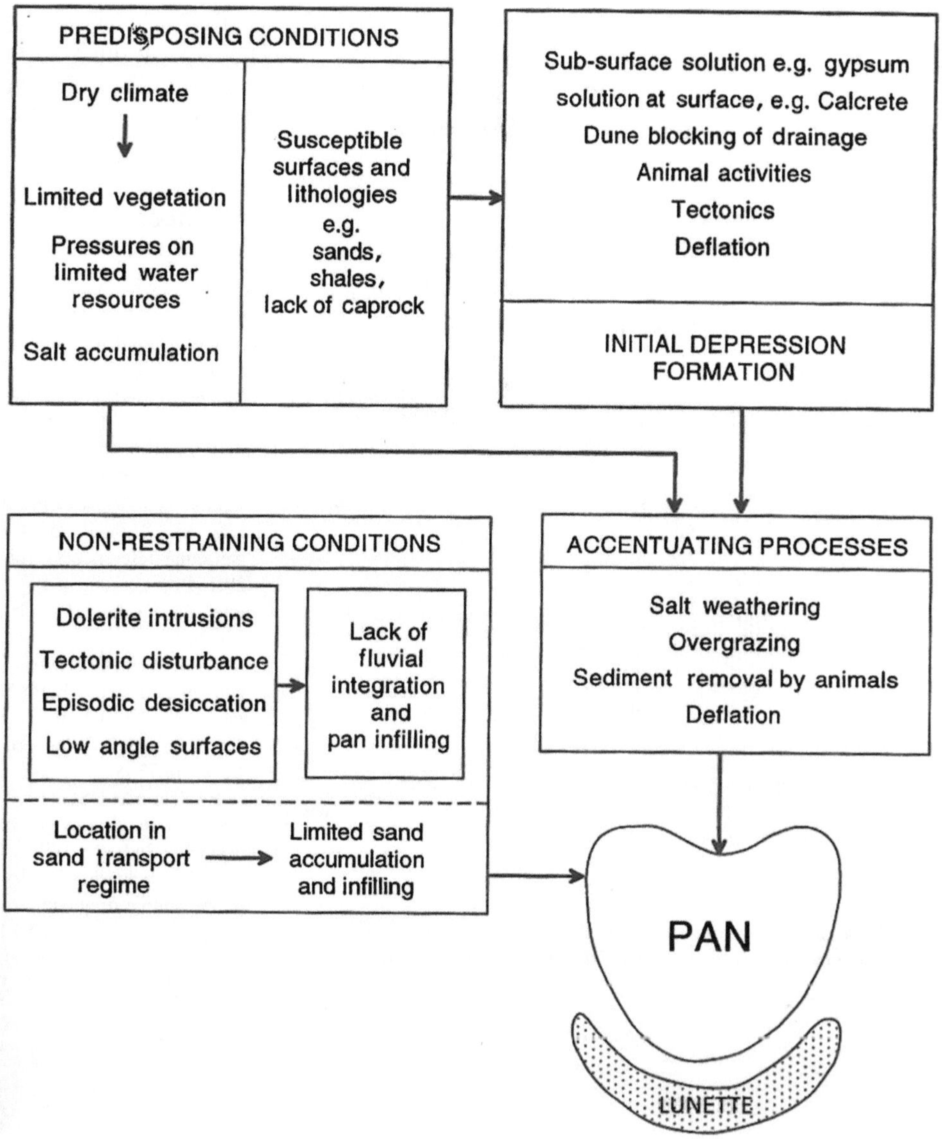

Fig. 23.3 A model of pan development (From Goudie 1999, Fig. 8.6)

which might cause infilling of an existing hollow, though pans can and do develop in inter-dune depressions, particularly in linear and parabolic dune fields.

Lunette dunes of the type found downwind from Koes (Fig. 23.4) were first named about seven decades ago in Australia, and since then have been recorded from many other semi-arid areas, including the High Plains of America, Tunisia, the Pampas of Argentina, the West Siberian Steppes, and the interior of South Africa (Goudie and Thomas 1985). They are composed of mixtures of clay, silt, sand and salts derived from the pan floor, and there may be several generations evident. At Koes, Lancaster (2000)

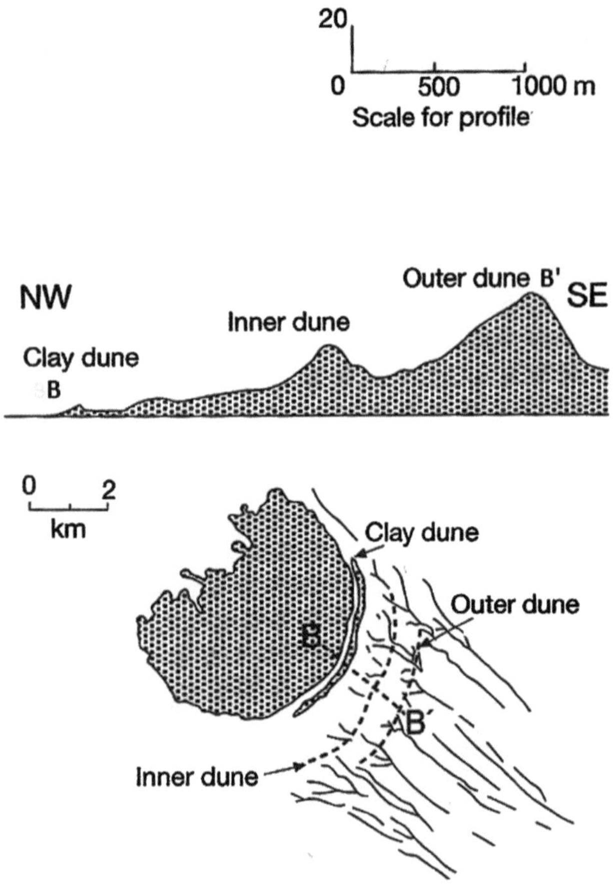

Fig. 23.4 Koes Pan cross-profile and sketch showing the three lunettes and the leeward linear dunes (Modified after Lancaster 2000)

identified an inner clay lunette and two outer, sandier lunettes. Linear dunes in the vicinity of Koes and its neighbours may have been active in the early Holocene (Blümel et al. 1998).

References

Alison MS (1899) On the origin and formation of pans. Trans Geol Soc South Afr 4:159–161

Blümel WD, Eitel B, Lang A (1998) Dunes in southeastern Namibia: evidence from Holocene environmental changes in the southwestern Kalahari based on thermoluminescence data. Palaeogeogr Palaeoclim Palaeoecol 138:139–149

Goudie AS (1999) Wind erosional landforms: yardangs and pans. In: Goudie AS, Livingstone I, Stokes S (eds) Aeolian environments, sediments and landforms. Wiley, Chichester, pp 167–180

Goudie AS, Thomas DSG (1985) Pans in southern Africa with particular reference to South Africa and Zimbabwe. Z Geomorph 29:1–19

Goudie AS, Viles HA (1997) Salt weathering hazards. Wiley, Chichester

Goudie AS, Wells GL (1995) The nature, distribution and formation of pans in arid zones. Earth-Sci Rev 38:1–69

Lancaster N (2000) Eolian deposits. In: Partridge TC, Maud R (eds) The Cenozoic of southern Africa. Oxford University Press, New York, pp 73–87

Shaw PA, Thomas DSG (1997) Pans, playas and salt lakes. In: Thomas DSG (ed) Arid zone geomorphology: process, form and change in drylands, 2nd edn. Wiley, Chichester, pp 293–317

Vickery KJ, Eckardt FD, Bryant RG (2013) A sub-basin scale dust plume source frequency inventory for southern Africa, 2005–2008. Geophys Res Lett. doi:10.1002/grl.50968

Circular Geomorphic Features: Roter Kamm and Gross Brukkaros 24

Abstract

Roter Kamm and Gross Brukkaros are two circular structures which at first sight might appear to have a similar origin. This is not the case. Roter Kamm, which lies in the Sperrgebiet, developed in 1,200 million year old Precambrian granitic rocks as a result of meteorite impact in the Pliocene c 3.7–5 million years ago. Gross Brukkaros is an impressive and isolated hill near Berseba, whose formation began towards the end of the Cretaceous c 75 million years ago. It started with the intrusion of carbonatite-rich magma into the Nama sediments that then covered this part of Namibia. The Brukkaros inselberg has since been brought into landscape prominence by downwearing of the surrounding surface.

24.1 Roter Kamm and Other Meteorite Impacts

In southern Namibia there are two circular structures which at first sight might appear to have a similar origin. These are Roter Kamm (Miller 2008, Chap. 26) and Gross Brukkaros (Miller 2008, Chap. 20). In reality there origins are very different.

Namibia is notable for the meteorites that have fallen onto its surface. In the north of the country, near Grootfontein, lies the Hoba Meteorite, a 60 ton iron-rich monster that is believed to be the largest in the world (Spencer 1932). It must have fallen to Earth quite slowly, as it failed to create an impact crater. In the south of the country, near Gibeon, covering an area of c 275 × 100 km, is one of the world's largest concentrations of meteorites. Also of importance is a beautiful circular crater called Roter Kamm (meaning 'red ridge' in German), which lies in the Sperrgebiet 50 km west of the mining town of Rosh Pinah at 27°46′S, 16°18′E. Its rim rises c 140 m above the surrounding plain and 160 m above the crater floor. The diameter from rim to rim is c 2.5 km. It has developed in 1,200 million year old Precambrian granitic rocks as a result of meteorite impact (Dietz 1965; Fudali 1973). It is thought to have formed in the Pliocene c 3.7–5 million years ago (Miller 2010). Its impact origin is indicated by the fact that it has a relatively low, narrow rim, a broad floor and *almost* perfect circularity. In fact it is not perfectly circular, for the NW–SE diameter is almost 250 m longer that its NE–SW diameter. This may be because the trajectory of the impacting body was north-westerly (Miller 2010). It also contains shock-metamorphosed rocks (Hecht et al. 2008) and major ejecta aprons have been found, especially on the northwest side of the crater. There are no volcanic rocks exposed and no evidence of a volcanic explosion. Equally, no remnants of the original meteorite that created the crater have been recovered. It lies within a zone of aeolian sands of the Sossus Sand Formation and has been seen as a possible analogue for such features on Mars. The sand obscures part of the rim and fills up a large portion of the crater by up to 500 m (Grant et al. 1997). The morphology of the dunes around and in the crater are shown in the Aster image (Fig. 24.1). To the north west of the crater there is an area that has been scooped out by deflation (Miller 2010). Lorenz et al. (2014) have suggested that the interactions between the crater and these eolian features provides a possible analogue for Titan.

A. Goudie and H. Viles, *Landscapes and Landforms of Namibia*, World Geomorphological Landscapes,
DOI 10.1007/978-94-017-8020-9_24,

Fig. 24.1 Aster image of Roter Kamm, showing the neighbouring sand dunes (courtesy of NASA)

Roter Kamm is one of a number of impact craters that has been recognised in southern Africa. Others include the ancient and enormous Vredefort structure, some 300 km across, Morokweng, Kalkkop and Tswaing (formerly called the Pretoria Salt Pan) all in South Africa and Kgagodi in Botswana.

24.2 Gross Brukkaros: Product of Differential Erosion

Rising up by about 600 m above the generally flat terrain between Mariental and Keetmanshoop in the Karas region of southern Namibia, and located about 15 km northeast of the primarily Nama town of Berseba, is an impressive and isolated hill called Gross Brukkaros. It Nama name is Geitsi Gubib. Lying c 40 km west of Tses at c 28°52′S and 17°47′ E) it has a basal diameter of c 7 km, and a steep-sided ring-shaped ridge bordering a central depression or crater and with a diameter of c 3 km. Its top reaches an altitude of 1,586 m above sea level. Brukkaros Mountain rests upon flat-lying reddish-brown sandstones and shales of the Fish River Subgroup (upper Nama Group; c 530 million years old), which were overlain by tillites and shales of the Dwyka Formation (Karoo Sequence; c 220 million years old). During subsequent uplift of southern African most of the Dwyka beds were removed by erosion, but a few remnants occur locally on the eastern and southwestern slopes of the mountain.

At first sight it might be construed to be a meteorite impact crater or volcano (Rogers 1915; Janse 1969), but the reality is different (Stachel et al. 1994, 1995). The formation of Gross Brukkaros (Fig. 24.2) began towards the end of the Cretaceous c 75 million years ago. It started with the intrusion of carbonatite-rich magma into the Nama sediments. This magma, working with superheated steam

Fig. 24.2 Google Earth image of Brukkaros. Scale bar 2 km. (© 2012 CNES/Spot Image, Digital Globe, Google)

derived from groundwater, caused the surface to bulge up and formed a 400 m high and 10 km wide dome, into which more magma was intruded, producing more steam. This then caused a great explosion that blew out the centre of the dome and created a crater. Groundwater drained into the new crater, where it came into contact with more magma, leading to further explosions from deeper levels within the Earth's crust. In the final stages of explosive activity material from 2 km deep was blasted out of the crater. This combined activity of water and magma is a process called phreato-magmatism. The hill is surrounded by many dikes and carbonatite pipes (Kurszlaukis and Lorenz 1997). Subsequently the crater was occupied by a lake, in which at least 300 m of sediments accumulated. These contain fossils that indicate the existence of a coniferous forest. Alluvial fans also accumulated. Subsequently, the Brukkaros inselberg has been brought into landscape prominence by downwearing of the surrounding surface—possibly an etchplain—since the late Cretaceous (Stengel and Busche 2002).

References

Dietz RS (1965) Roter Kamm, Southwest Africa: probable meteorite crater. Meteoritics 2:311–314

Fudali RF (1973) Roter Kamm: evidence for an impact origin. Meteoritics 8:245–257

Grant JA, Koeberl C, Reimold WU, Schultz PH (1997) Gradation of the Roter Kamm impact crater, Namibia. J Geophys Res 102 (E7):16327–16338

Hecht L, Reimold WU, Sherlock S, Tagle R, Koeberl C, Schmitt R-T (2008) New impact-melt rock from the Roter Kamm impact structure, Namibia: further constraints on impact age, melt rock chemistry, and projectile composition. Meteoritics Plan Sci 43:1201–1218

Janse AJA (1969) Gross Brukkaros, a probable carbonatite volcano in the Nama plateau of Southwest Africa. Geol Soc Am Bull 80:573–586

Kurszlaukis S, Lorenz V (1997) Volcanological features of a low-viscosity melt: the carbonatitic Gross Brukkaros volcanic field, Namibia. Bull Volcanol 58:421–431

Lorenz RD, Miller RM, Hayes AG, Radebaugh J, Barnes JW, Arnold KD, Chandler CK (2014) Roter Kamm impact structure, Namibia: new field and remote sensing insights and analogies with Titan Dune: Crater interactions. In: Lunar and Planetary Science Conference Abstracts, vol 45, p 1479

Miller RMG (2008) The geology of Namibia, vol 3. Geological Survey of Namibia, Windhoek

Miller RMG (2010) Roter Kamm impact crater of Namibia: new data on rim structure, target rock geochemistry, ejecta, and meteorite trajectory. Geol Soc Am Spec Pap 465:489–508

Rogers AW (1915) Geitzi Gubib, an old volcano. Trans Roy Soc S Afr 5:247–258

Spencer LJ (1932) Hoba (South–West Africa), the largest known meteorite. Mineral Mag 23:1–19

Stachel T, Lorenz V, Stanistreet IG (1994) Gross Brukkaros (Namibia): an enigmatic crater-fill reinterpreted as due to Cretaceous caldera evolution. Bull Volcanol 56:386–397

Stachel T, Brey G, Lorenz V (1995) Carbonatite magmatism and fenitization of the epiclastic caldera-fill at Gross Brukkaros (Namibia). Bull Volcanol 57:185–196

Stengel I, Busche D (2002) Namibia's oldest inselbergs—exhumed landforms of a Late Proterozoic Etchplain. In: Yang X (ed) Desert and alpine environments. China Ocean Press, Beijing, pp 121–143

25 Mounds, Rings and Stripes

Abstract

There are various types of patterned ground in Namibia. The circular 'Fairy Rings' of the pro-Namib are a vegetation pattern that has been likened to an ostrich skin. Mounds are also abundant. In the Western Cape in South Africa they are termed 'heuweltjies' and are thought to resemble the mima mounds of North America. Further north in Namibia one moves into an environment where large numbers of mounds are produced by the termite, *Macrotermes*. In the southern Kalahari there are extensive areas of banded vegetation stripes, which are generally called tiger bush or brousse tigreé. The possible origins of these different patterns are described.

Namibia is rich in patterned ground. There is a range of mounds, rings and stripes. They are spread over much of the country and so no one site has been selected to visit and view them, but fairy circles are particularly clearly seen in Hartmann's Valley in the north west, mounds are commonly found in the north east of the country, and banded vegetation on the west and south eastern margins of the Weissrand Plateau.

The circular 'Fairy Circles' of the pro-Namib (a term used to describe the eastern, slightly wetter margins of the Namib Desert) in Namibia (Albrecht et al. 2001) are a type of intriguing, but little understood, vegetation pattern, that has been likened to an ostrich skin (Picker 2012) (Fig. 25.1). They occur in their thousands in areas with sandy soils and where the mean annual rainfall is 50–150 mm. They consist of circular bare areas c 2–12 m in diameter surrounded by perennial grasses such as *Stipagrostis*. In general they seem to decrease in size from north to south, with the largest occurring in Hartmann's Valley in the far north of the country. Here there also appear to be some areas where the circles have 'healed', leaving circular patches of vegetation (Fig. 25.2a). Other features in that area are of interest because they appear to have been elongated and possibly deflated by south westerly winds (Fig. 25.2b). Fairy circles are widespread between the Orange River and southern Angola (Becker and Getzin 2000; Picker et al. 2012). It is possible that their origin owes something to the foraging action of termites (Grube 2002; Juergens 2013) or ants (Picker et al. 2012) or to growth inhibition as a result of allelopathic compounds released by dead *Euphorbia damarana* plants, but as yet there is no entirely satisfactory explanation for their origin (van Rooyen et al. 2004). It has also been proposed that they may be the result of micro-seepage of gases and hydrocarbons (Naudé et al. 2011). They are ephemeral features that have a lifespan on the order of decades (Tschinkel 2012).

Mounds are also abundant (Fig. 25.3). In the winter rainfall belt of the Western Cape in South Africa they are termed 'heuweltjies' and are thought to resemble the enigmatic mima mounds of North America (Francis et al. 2012). In South Africa they average 17 m in diameter and 1.45 m in height (Picker et al. 2007). Their average density is just under 300 mounds per km^2. Examination of Google Earth images suggests that morphologically identical features extend in south central Namibia north of the Orange River in a belt about 30 km wide up to c 27–26°S.

They have been the subject of considerable debate as to origin, with the role of bioturbation by rodents being favoured for the analogous mima mounds of the USA (Burnham and Johnson 2012; Gabet et al. 2014). This mechanism, involving mole rats, has been advocated for examples from the Cape Region in South Africa (Cox et al. 1987), but in southern Africa the southern harvester termite, *Microhodotermes viator*, also appears to be implicated. On the other hand, Cramer et al. (2012) have suggested that the mounds are the result of the protection of the

A. Goudie and H. Viles, *Landscapes and Landforms of Namibia*, World Geomorphological Landscapes,
DOI 10.1007/978-94-017-8020-9_25,

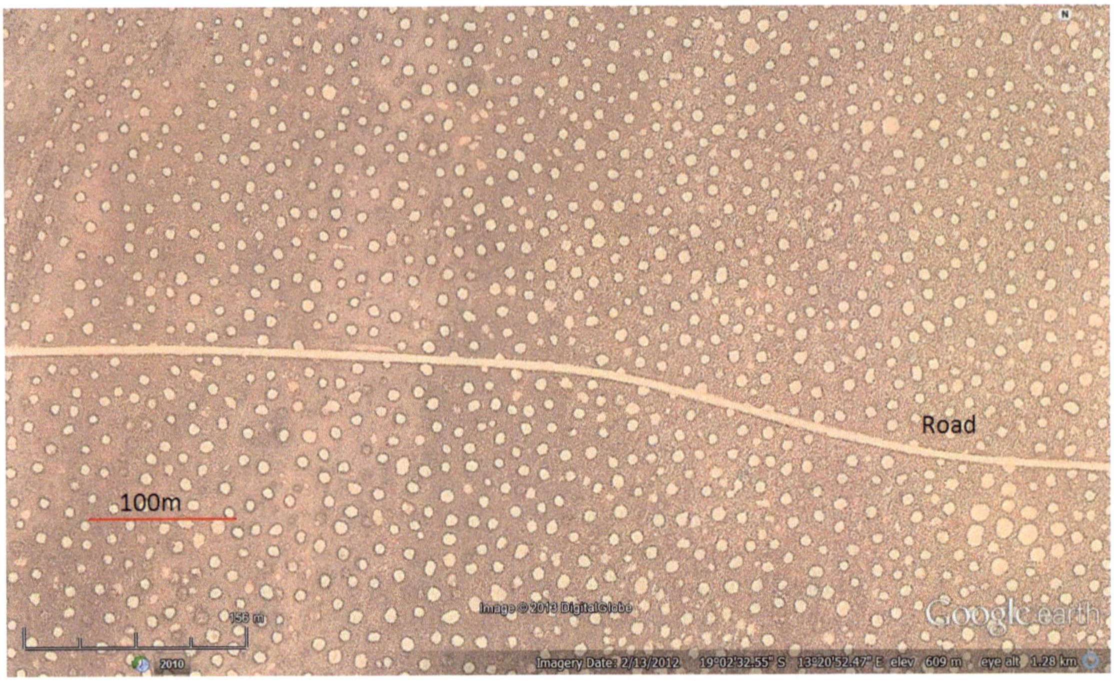

Fig. 25.1 Google Earth image of fairy circles. Scale bar is 100 m. (© 2013 Digital Globe)

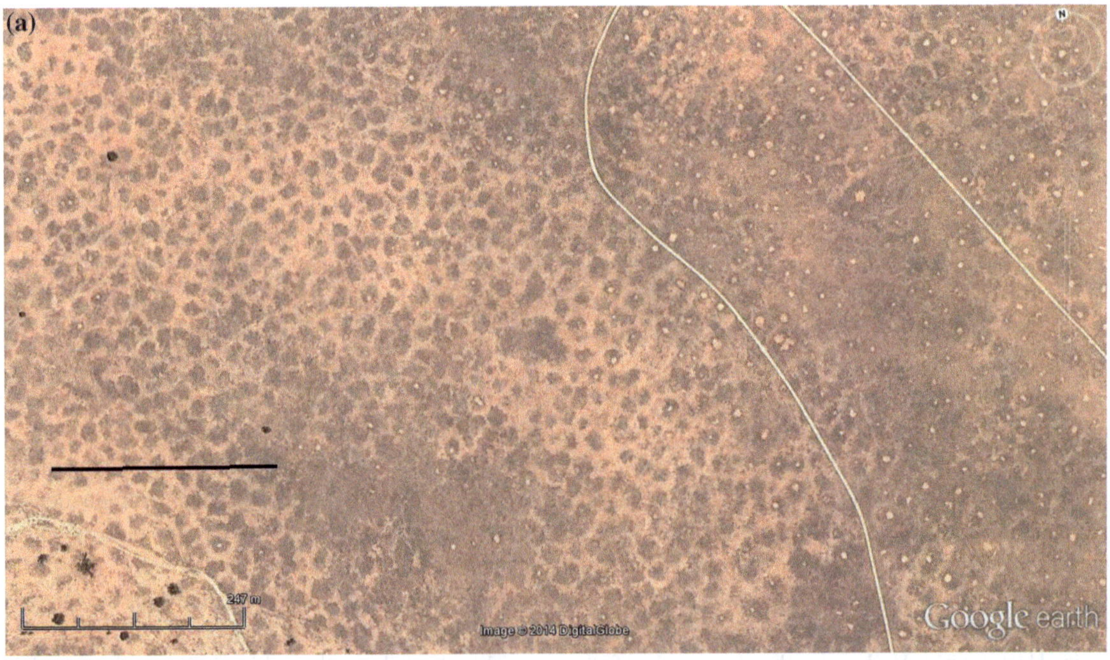

Fig. 25.2 **a** Google Earth image of possible 'healed' fairy circles in the west of the Hartmann's valley. On the east side of the image there are some partially healed forms. Scale bar, 0.25 km. (© 2014 Digital Globe). **b** Google Earth image of elongated fairy circles from the far north of Namibia. Scale bar 0.2 km. (© 2012 GeoEye)

soil from erosion by the presence of roots of regularly spaced woody shrubs, and that this protection produced deeper soils that favoured secondary faunal activity, including that by termites.

Further north in Namibia one moves into an environment where large numbers of another sort of mound are produced by another type of termite, *Macrotermes* (Turner et al. 2006; Grohmann et al. 2010) (Fig. 25.4). The morphology of these

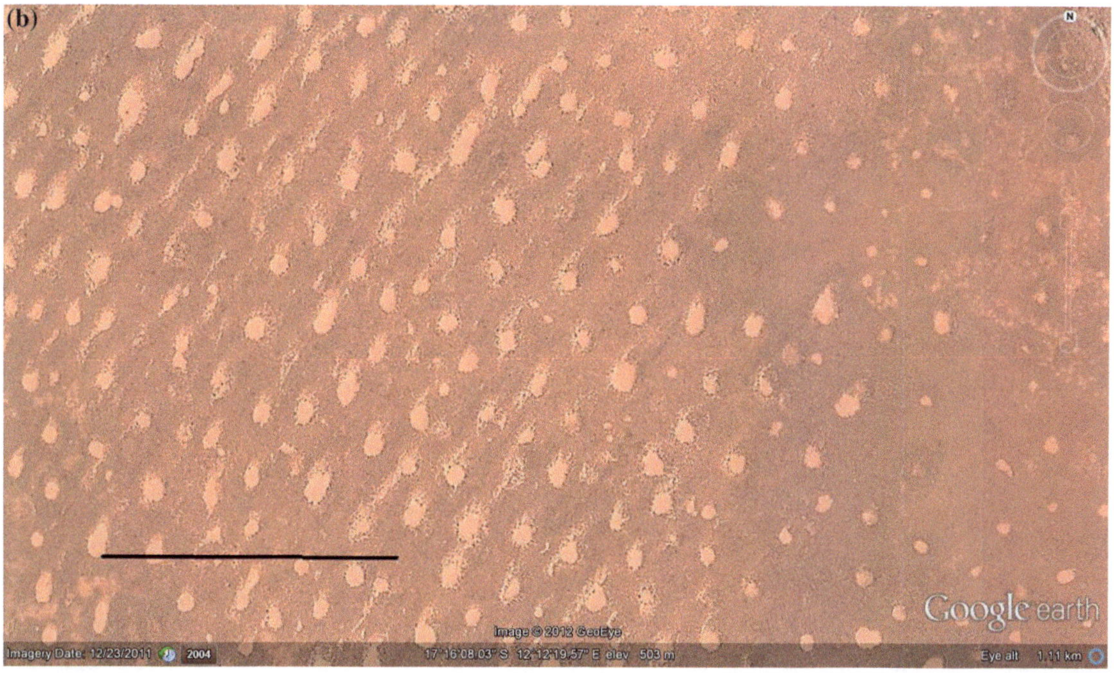

Fig. 25.2 continued

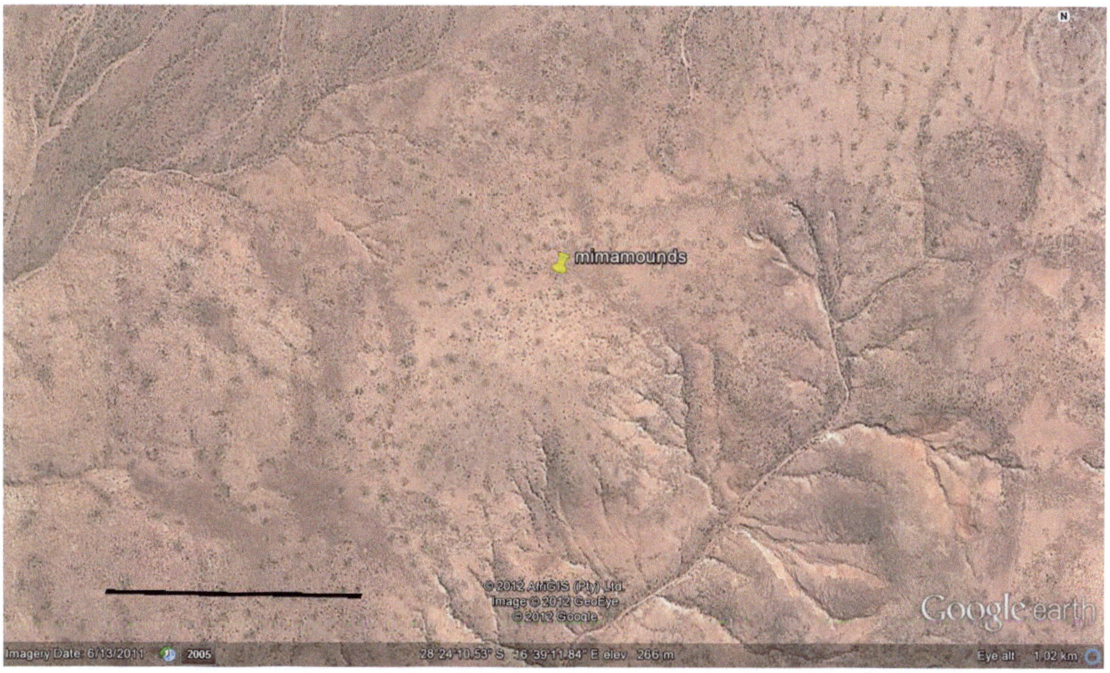

Fig. 25.3 Google Earth image of heuweltjie mounds. Scale bar 0.2 km. (2012 ©AfriGIS (Pty) Ltd, GeoEye, Google)

mounds consists of three components: a central cone-shaped mound with an average basal circumference of c 7.5 m, a tall thin spire which tilts northward at an angle similar to the sun's average zenith angle, and a broad outwash pediment that results from erosion of the mound (Turner 2000).

In the southern parts of Namibia there are extensive areas of banded vegetation stripes (Fig. 25.5), which are generally called tiger bush or brousse tigrée, but have only been very briefly described in Namibia by Stengel (2000) and by Goudie (2007). Banded vegetation (tiger bush or brousse

Fig. 25.4 Google Earth image of termite mounds in Caprivi. Scale bar is 0.5 km. (2012 © Google, GeoEye, US Dept of State Geographer)

Fig. 25.5 Google Earth image of tiger bush stripes near Bullsport. Scale bar 1 km. (2012 © GeoEye, Google)

tigrée) is a clear vegetational pattern when observed from air in arid and semiarid areas of the entire planet. It gives rise to a peculiar landscape formed by alternating *bands* of vegetation (grass, shrubs or trees) and *interbands* of bare soil (sometimes with termite mounds and or clumps of grasses). The vegetated bands are usually curved, with the convex part facing upslope. The bands can form long and continuous stripes of vegetation or they can be interrupted and 'degraded' to different degrees (Valentin et al. 1999). Indeed, four different patterns of vegetation have been defined by d'Herbes et al. (2001): *banded, fuzzy, dashed or dotted,* and *spotted*. The principal environmental controlling factors

Fig. 25.6 Landsat image of inter-dunal tiger bush (Courtesy of NASA)

determining the different patterns are two: slope gradient (very low) and mean annual rainfall (arid or semiarid regimes). Such phenomena are extensively developed in the world's drylands where there is between c 100 and c 600 mm of mean annual rainfall (Tongway and Ludwig 2001).

Banded and patterned vegetation is thought to be the best adaptation, for selected species, to water harvesting in arid and semiarid environment as a result of spatial self-organization (Rietkerk et al. 2002), with erratic and extremely variable rainfall. Banded patterns result from the interplay between mechanisms of short range facilitation and of long range competition with surrounding vegetation (D'Odorico et al. 2006), even if there is not yet a generally accepted model of banding formation and new analysis and models are proposed with increased frequency (see, for example, Ursino 2007). The general mechanism invoked can be summarized as follows: when water flows over the bare soil areas (interband) it does not infiltrate due to the presence of crusty, almost impermeable, muddy soils. As soon as the water sheet reaches the vegetated bands it passes over a very porous soil (due to the disintegrating action of the roots) and it infiltrates downward. The water act also as a conveyor sheet and leads to further accumulation of litter, sand, and other organic and inorganic particles at the vegetated band that acts thus as a dam.

Tiger bush occurs in two isolated zones near Bullsport (24°09′S; 16°25′E) and Mara (24°50′S; 16°38′E) on the edge of the Naukluft Mountains, but the main zone occurs between latitudes S24° and S28° to the west and south east of the Weissrand Plateau between Marienthal, Keetmanshoop and the South African border. It has developed where the mean annual rainfall averages between 150 and 200 mm, but does not appear to occur in drier areas than this, probably because there is less sheetflood activity and a much sparser vegetation cover. Examples of tiger bush are present in three main geomorphological situations: in interdunal corridors (Fig. 25.6), on low angle slopes developed on alluvium or outcrops of fine-grained Ecca and Dwyka sedimentary rocks, and along the courses of ephemeral streams. They do not occur on areas with complete aeolian sand cover (probably because this restricts overland flow), on the top of the Weissrand Plateau (where drainage mostly seeps underground through the aligned karstic hollows or dayas), or on steep slopes in mountainous areas, where runoff tends to be more concentrated. The dark stripes tend to occur with a frequency of around 4–10 per km (i.e. the wavelength of a cycle including a band and an interband is around 100–250 m). This is broadly comparable to figures reported from other areas in Africa and in Australia (see, for example, Valentin and Poesen 1999, Table 2, p. 7). Another tiger bush location is on the granite pediment slopes near to Spitzkoppe (Fig. 25.7).

Finally, some linear vegetation patterns, found near Berseba and Tses in southern Namibia, result from structures in the underlying bedrock (Stengel and Busche 2003), and

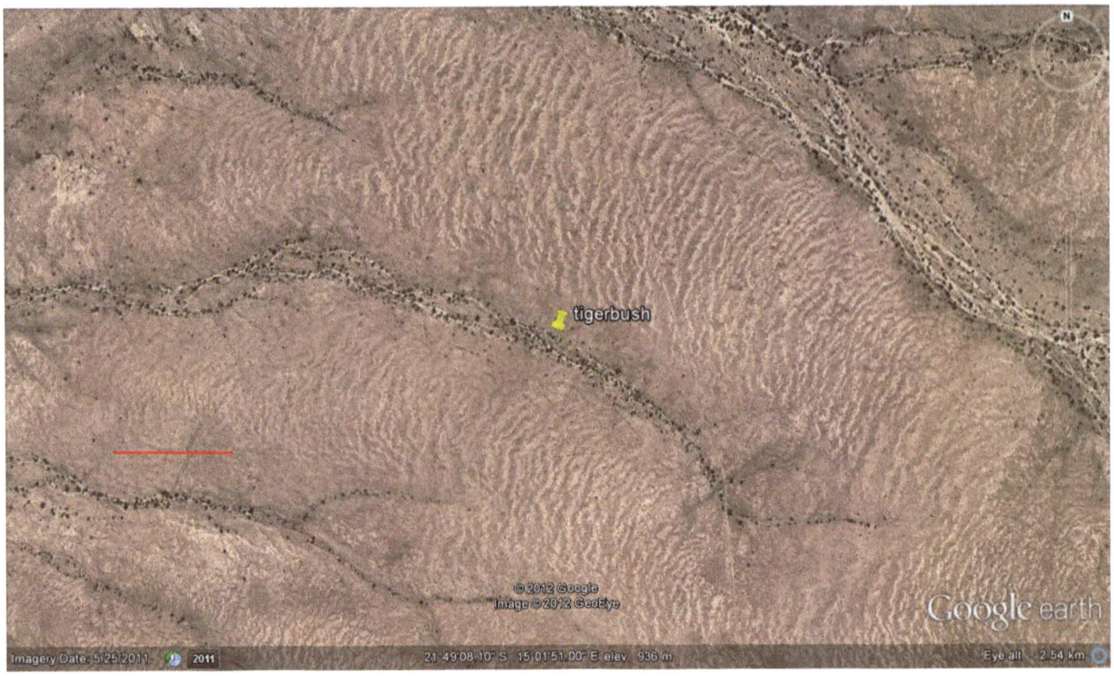

Fig. 25.7 Google Earth image of tiger bush on the granitic pediments south west of Spitzkoppe. (© 2012 Google Image, GeoEye)

polygonal vegetation patterns of uncertain origin have been found in calcretes and Tsondab Sandstone in the Kuiseb terraces and interdunes to the south of Gobabeb (Ollier and Seely 1977).

References

Albrecht CF, Joubert JJ, De Rycke PH (2001) Origin of the enigmatic, circular, barren patches ('fairy rings') of the pro-Namib. S Afr J Sci 97:23–27

Becker T, Getzin S (2000) The fairy circles of Kaokoland (North–West Namibia): origin, distribution, and characteristics. Basic Appl Ecol 1:149–159

Burnham JHL, Johnson DL (eds) (2012) Mima mounds: the case for polygenesis and bioturbation. Geol Soc Am Spec Pub, p 490

Cox GW, Lovegrove BG, Siegfried WR (1987) The small stone content of mima-like mounds in the South African Cape region: implications for mound origin. Catena 14:165–176

Cramer MD, Innes SN, Midgley JJ (2012) Hard evidence that *heuweltjie* earth mounds are relictual features produced by differential erosion. Palaeogeogr Palaeoclimatol Palaeoecol 350–352:189–197

d'Herbes J-M, Valentin C, Tongway DJ, Leprun J-C (2001) Banded vegetation patterns and related structures. In: Tongway DJ, Valentin C, Seghieri J (eds) Banded vegetation patterning in arid and semiarid environments: ecological processes and consequences for management. Ecological studies no. 149. Springer, New York, pp 1–19

D'Odorico P, Laio F, Ridolfi L (2006) Patterns as indicators of productivity enhancement by facilitation and competition in dryland vegetation. J Geophys Res 111:G03010

Francis ML, Ellis F, Lambrechts JJN, Poch RM (2012) A micromorphological view through a Namaqualand termitaria (*Heuweltjie*, a mima-like mound). Catena 100:57–73

Gabet EJ, Perron JT, Johnson DL (2014) Biotic origin for mima mounds supported by numerical modeling. Geomorphology 206:58–66

Goudie AS (2007) Desert landforms in Namibia: a Landsat interpretation. In: Goudie AS, Kalvoda J (eds) Geomorphological variations. Nakladatelstvi P3K, Prague, pp 19–36

Grohmann C, Oldeland J, Stoyan D, Linsenmair KE (2010) Multi-scale pattern analysis of a mound-building termite species. Insectes Soc 57:477–486

Grube S (2002) The fairy circles of Kaokoland (Northwest Namibia)—is the harvester termite Hodotermes mossambicus the prime causal factor in circle formation? Basic Appl Ecol 3:367–370

Juergens N (2013) The biological underpinnings of Namib desert fairy circles. Science 339:1618–1621

Naudé Y, van Rooyen MW, Rohwer ER (2011) Evidence for a geochemical origin of the mysterious circles in the Pro-Namib desert. J Arid Env 75:446–456

Ollier CD, Seely MK (1977) Patterned ground near Gobabeb, central Namib desert. Madoqua 10:213–214

Picker M (2012) Little landscapers: social insects transform landscapes by recycling and releasing nutrients and increasing floral diversity. Veld Flora 98:174–177

Picker MD, Hoffman MT, Leverton B (2007) Density of *Microhodotermes viator* (Hodotermitidae) mounds in southern Africa in relation to rainfall and vegetative productivity gradients. J Zool 271:37–44

Picker MD, Ross-Gillespie V, Vlieghe K, Moll E (2012) Ants and the enigmatic Namibia fairy circles—cause and effect? Ecol Entomol 37:33–42

Rietkerk M, Boerlijst MC, van Langevelde F, HilleRisLambers R, van de Koppel J, Kumar L, Prins HHT, de Roos AM (2002) Self-organization of vegetation in arid ecosystems. Am Nat, Notes Comments 160:524–530

van Rooyen MW, Theron GK, van Rooyen N, Jankowitz WJ, Matthews WS (2004) Mysterious circles in the Namib desert: review of hypotheses on their origin. J Arid Env 57:467–485

Stengel I (2000) Brousse tigreé – Muster und Materialverspülung und Pflanzewuchs in Südnamibia. Petermanns Geogr Mitt 144(2):4–5

Stengel I, Busche D (2003) Structurally induced sand ridge and vegetation patterns in the Berseba/Tses area, southern Namibia. Geo-Öko 24:163–176
Tongway DJ, Ludwig JA (2001) Theories on the origins, maintenance, dynamics, and functioning of banded landscapes. Ecol Stud 149:20–31
Tschinkel WR (2012) The life cycle and life span of Namibian fairy circles. PLoS ONE 7(6):e38056
Turner JS (2000) Architecture and morphogenesis in the mound of Macrotermes michaelseni (Sjöstedt) (Isoptera: termitidae, Macrotermitinae) in northern Namibia. Cimbebasia 16:143–175
Turner S, Marais E, Vinte M, Mudemgi A, Park W (2006) Termites, water and soils. Agricola 16:40–45
Ursino N (2007) Modeling banded vegetation patterns in semiarid regions: interdependence between biomass growth rate and relevant hydrological processes. Water Res Res 43:W04412
Valentin C, d'Herbès J-M, Poesen J (1999) Soil and water components of banded vegetation patterns. Catena 37:1–24
Valentin C, Poesen J (1999) The significance of soil, water and landscape processes in banded vegetation patterning. Catena 37:1–273

Part III

Heritage Issues

Valuing, Conserving and Raising Awareness of Namibia's Landscapes and Landforms

26

Abstract

The geomorphological landscapes of Namibia are of huge importance within their own right and also for the foundations they provide for biodiversity. Whilst many landscapes are conserved with statutory designations, other planned conservation zones have not yet been implemented. Many human activities threaten the integrity of Namibia's geomorphological landscapes, and conflicts between economic development and geoconservation are not always easy to resolve. Several bodies, including the Geological Survey of Namibia, are working hard to improve public understanding of the special and unique geomorphological landscape of the country. The importance of geotourism for sustaining Namibia's geomorphological landscapes should not be underestimated.

26.1 Introduction

The landscapes and landforms of Namibia are a highly important natural resource for the country, matching in diversity and touristic value the better-documented ecosystems and species. Whilst people undoubtedly flock to Namibia to see elephants, lions and other charismatic species, the backdrop provided by the rocks and physical landscape diversity across the country is also of immense interest to most tourists. The sheer beauty of the Namib Sand Sea, as captured in so many tourist photos of the huge star dunes which surround Sossus Vlei, demonstrates the contribution that geomorphological landscapes make to inspiring tourists in Namibia. Valuing, conserving and marketing the geomorphological landscapes of Namibia are three key tasks in order to maximise ensure their sustainable conservation and enhance sustainable development.

26.2 Valuing Geomorphological Landscapes in Namibia: Geodiversity and Geoheritage

As reviewed by Gray (2013) geodiversity is an increasingly used concept to describe the variety of geology, geomorphology and soils within an area. There are a range of different metrics which can be used to measure such geodiversity (such as topographical diversity or relief variation, number of rock types, number of iconic geosites), but whichever one is used, Namibia would score very highly. It is a truly geodiverse country and one in which, because of the relatively arid conditions and lack of thick vegetation cover, the geological and geomorphological dimensions of the landscape are clear to see. As Gray (2013) points out, geodiversity is a very important factor underpinning biodiversity. Without the rich geodiversity of Namibia, the biodiversity of the country would be impoverished. For example, the habitats provided by the ephemeral river systems of the Skeleton Coast are vital to the maintenance of the desert-adapted elephant population which roams widely across these river systems. The elephants are supported by the riparian forests which grow within and around the ephemeral stream channels, nurtured by the shallow groundwater and occasional surface flow. Furthermore, the lichen fields of the fog-influenced gravel plains along the northern Skeleton Coast are crucial parts of Namib's biodiversity (including many rare and endemic species) and rely on the desert pavement surfaces to provide a home.

As Thomas (2012) notes, geomorphological processes are, in turn, vital for the maintenance of geodiversity—without the interplay of denudation and tectonic activity geodiversity of Earth would have declined over millions of years. Geomorphic processes refresh the landscape, creating further elements of geodiversity. Gray (2013) also reminds

A. Goudie and H. Viles, *Landscapes and Landforms of Namibia*, World Geomorphological Landscapes,
DOI 10.1007/978-94-017-8020-9_26,

us that geodiversity is important in its own right, as testament to the Earth's history and as something that we should value for its beauty, cultural associations and scientific significance. Many of the landscapes reviewed in this book, such as Spitzkoppe and Brandberg are iconic parts of Namibia's geodiversity and geoheritage. However, smaller and less obviously dramatic landscapes such as those dominated by the salt weathering at Soutrivier and the tufas of the Naukluft Mountains are also important components of the complex and dynamic geodiversity of Namibia.

26.3 Conserving Geomorphological Landscapes in Namibia: From National Monuments to World Heritage Sites

The value of Namibia's geomorphological landscapes can partly be measured by the number of sites listed for conservation at local, national and international level (Fig. 26.1). Namibia's National Heritage Council catalogues, names and conserves sites of outstanding importance. These are designated as National Monuments. Among the geomorphological sites for which they have responsibility are Brandberg, the Fish River Canyon, Mukorob, Gaub Cave, Otjikoto, Spitzkoppe, Twyfelfontein, and Waterberg. Namibia has a large protected area network which consists of around 20 national parks and nature reserves. While much of this is essentially done for conservation of fauna and flora, these protected areas also, even if incidentally, give a large degree of protection to key geomorphological landscapes. They include the Namib Naukluft, the Waterberg Plateau, Etosha Pan, the Fish River Canyon, and almost the entire coastline. Within these conserved areas, strict rules apply to prevent damage to the landscape from, for example, off-road driving.

Namibia has two sites that have been given UNESCO World Heritage status, meaning that they have been acknowledged as being of outstanding universal value—Twyfelfontein (2007) and the Namib Sand Sea (2013). There are in addition additional sites that have been placed on the Tentative List: Brandberg National Monument Area (2002), the Fish River Canyon (2002), and the Welwitschia Plains between the Swakop and Khan rivers (2002). These tentative list sites all owe at least part of their value to the underlying geomorphology and geology.

Of the two current Namibian World Heritage sites, Twyfelfontein is listed only for its cultural heritage values, because of the rock art found there, but this rock art would not exist without the canvas provided by the rocks and geomorphological landscape (see Chap. 7). Undoubtedly, it is the award of World Heritage Status to the Namib Sand Sea (Fig. 26.2) that is the most important event in terms of geomorphological landscape protection with the country. The nomination file illustrates that the site meets all four of the natural criteria against which world heritage sites can be listed, including aesthetic, geological, geomorphological and biological.

Other attempts at conserving the geomorphological landscape have been less successful to date, but are nevertheless important and reflect the richness of the Namibian geology and geomorphology. In 2004, a proposal for the 'Gondwanaland Geopark' was produced by Gabi Schneider and colleagues with UNESCO funding. Geoparks are another UNESCO initiative designed to complement and extend the World Heritage list by protecting a larger number of geomorphologically and geologically important areas. What marks geoparks out as special is their joint commitments to conservation of geodiversity, encouragement of economic development and furthering sustainable development within each park. The Global Geoparks Network was established in 2004 and as of 2014 there are 100 Geoparks worldwide with high numbers in China and Europe. There are as yet no geoparks in Africa.

The Gondwanaland Geopark is an ambitious concept—covering an area of c 60,000 km^2 between 20 and 22°S and 13–16°E in the central western part of Namibia. The geomorphological contributions to the geopark are clarified early on in the proposal which notes that 'the area boasts scenic landforms… and bears witness to geomorphic processes' (Schneider and Schneider 2004, p. 1). It has been designed to contain several of the sites described in this book—including Brandberg, Spitzkoppe, Erongo and the Etendeka Plateau. Addressing the aims of geoparks to promote economic and sustainable development, the proposal documents how the c 45,000 population of the area could benefit. They foresee enhanced geotourism activities, alongside the long-established mineral extraction activities (e.g. for amethyst, dimension stone, salt and tungsten) and extensive farming systems which cover some parts of the proposed park. By 2014 the Gondwanaland Geopark had still not been turned into reality, largely because of delays in getting the legislative framework in place.

Conservation of geodiversity and geoheritage reflects not only the values of the geomorphological landscape, but is also a measure of the threats to it. Complex and dynamic geomorphic environments can be threatened by a range of human activities, such as bad land management practices and off-road driving (which can both accelerate soil erosion), mining (which can cause very large scars on the landscape, not only directly destroying landforms but also reducing the aesthetic value of the area) and groundwater extraction.

With regard to threats to Namibia's geomorphological heritage, considerable concern has been expressed with regard to the potential spread of mining activity, not least in the Central Namib (Wassenaar et al. 2013). Many areas, have, for example, been granted uranium mining licences (Fig. 26.3). Several licences have been granted within the proposed Gondwanaland Geopark. A considerable threat to

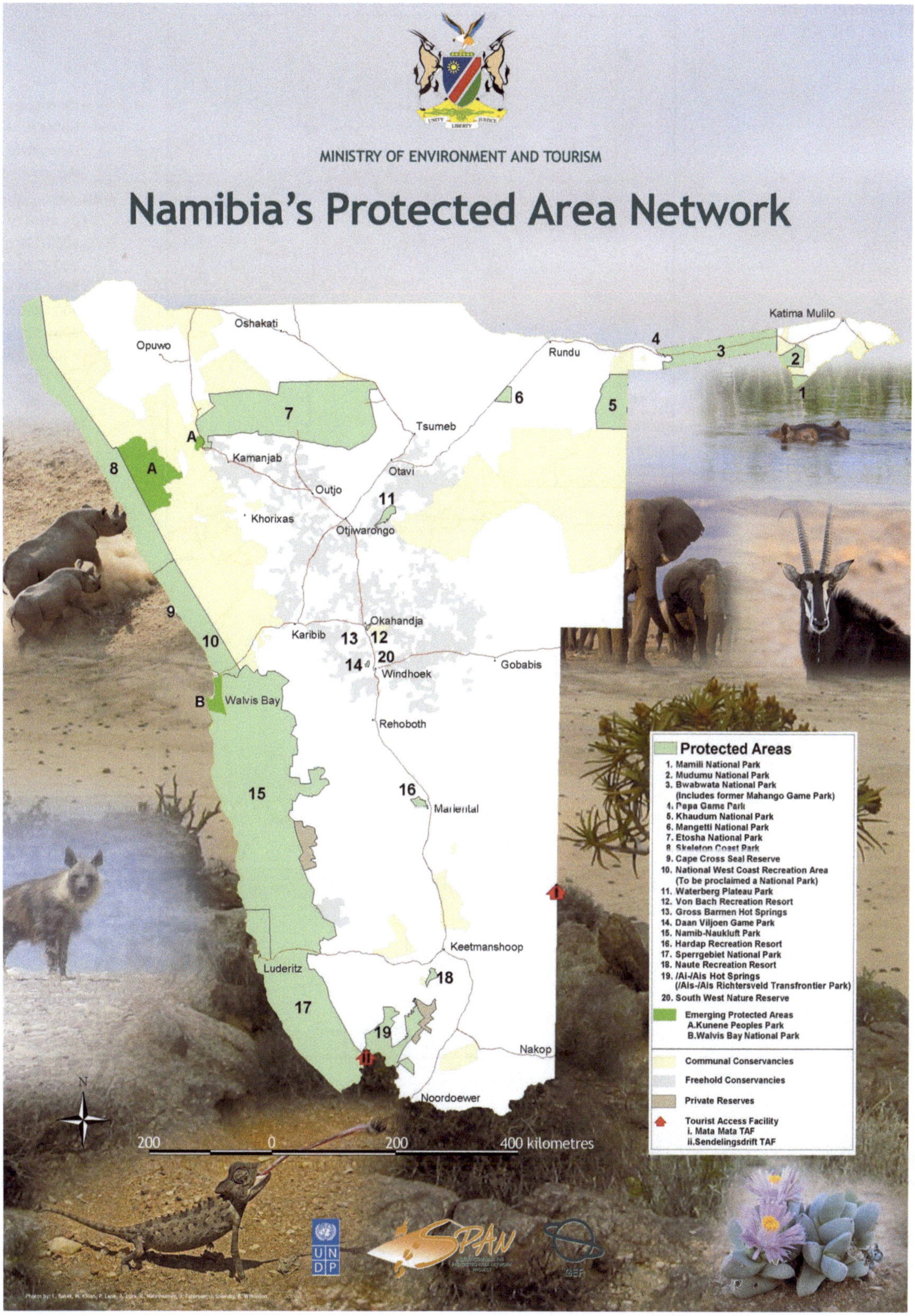

Fig. 26.1 Protected areas in Namibia (http://www.met.gov.na/Pages/Protectedareas.aspx) (accessed 14th February 2014)

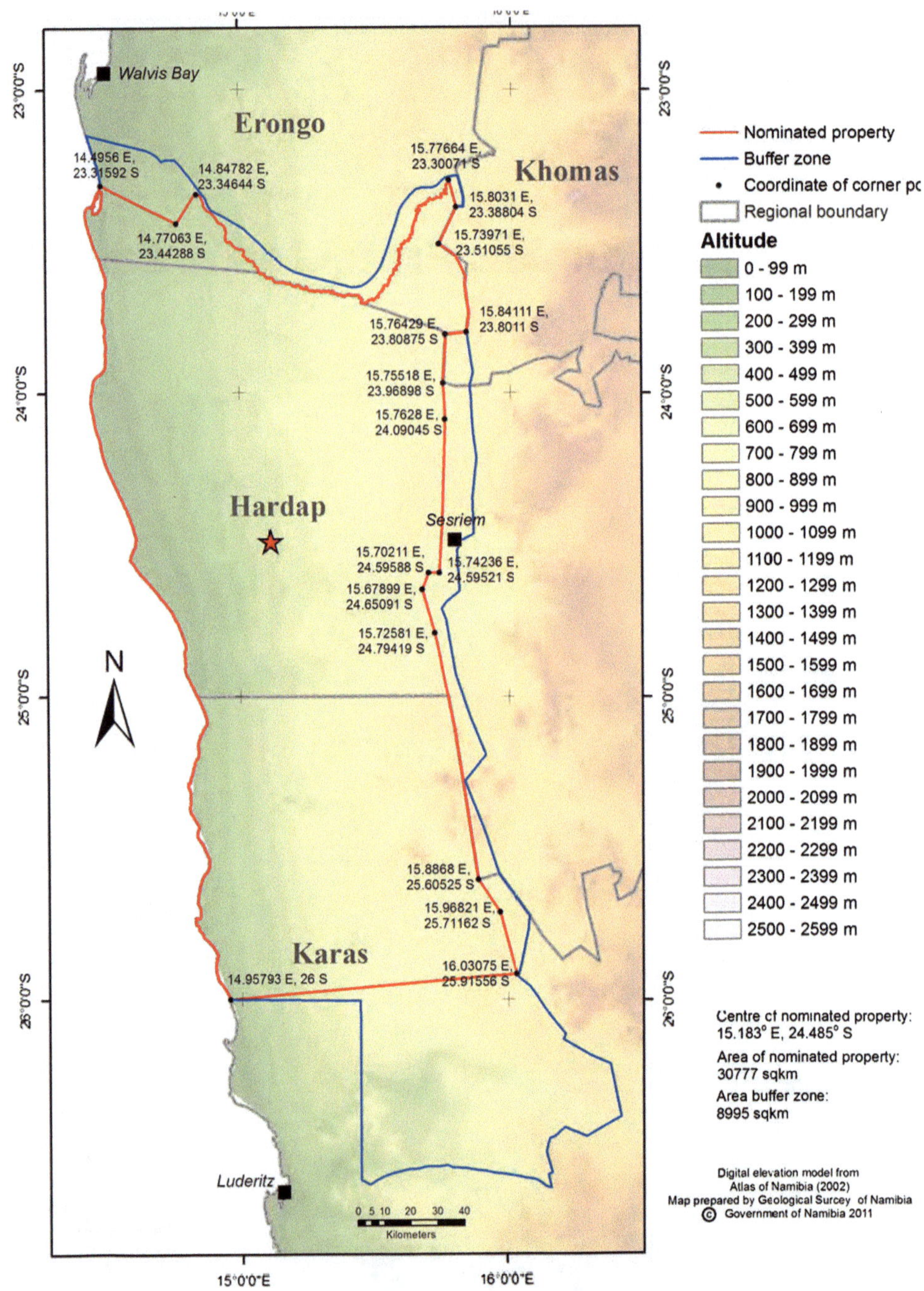

Fig. 26.2 Delimitation of the Namib sand sea UNESCO World Heritage Site

the Namib Desert comes from off-road driving associated with mineral prospecting and tourism. The impact is the greatest on the gravel plains where vehicle tracks may remain for more than 40 years because the rainfall is too episodic and sparse to erase them. These unsightly tracks cause long-lasting damage to lichen fields. Lichens are particularly sensitive to mechanical damage as they grow extremely slowly. A major threat to the Namib-Naukluft National Park is the drop in the water table along the Kuiseb River, caused to a large degree by the extraction of groundwater near Walvis Bay. The Kuiseb River and the vegetation within it act as a windbreak that retards the northwards movement of the Namib Sand Sea onto the gravel plains, and so any reduction in the river's surface flow or in the vegetation cover along its channel could have widespread consequences.

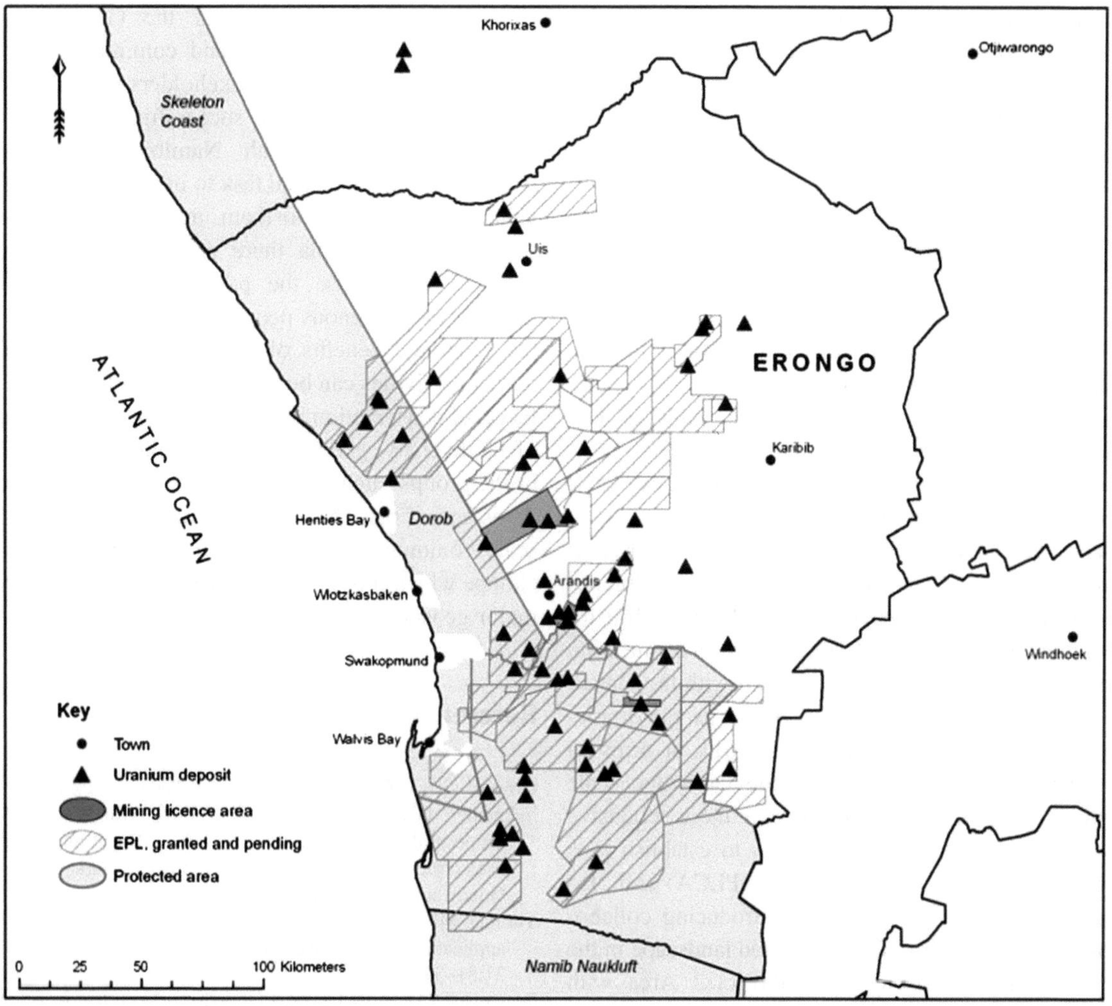

Fig. 26.3 Uranium deposits and mining licence areas. *EPL* is an exclusive prospecting licence (from Wassenaar et al. 2013, Fig. 2)

26.4 Raising Awareness of Namibia's Geomorphological Landscapes: Information and Geotourism

As has been found in many other areas, public understanding and appreciation of geomorphology and geology is much less developed than that of species and ecosystems. Whilst many Namibian people past and present have undoubtedly had close relationships with the physical landscape, there are few records of intangible cultural associations such as myths, told stories and legends. Establishing a collection of the diverse meanings given to different landscapes by different tribal and other groups would be of immense value. Modern scientific studies on the Namibian landscape, many of which have been reviewed in this book, are diverse and instructive—but not easy for lay people to access nor understand. However, in comparison with many countries within Western Europe, for example, the depth of scientific information about many even iconic parts of the Namibian landscape (such as the Fish River Canyon) is very limited. The Geological Survey of Namibia not only carries out a vast range of fundamental research on the country's landscapes and resources, but has also produced a range of informative posters about major geosites (e.g. Brukkaros, Twyfelfontein, Sperrgebiet, Gamsberg and Fish River) freely available from their website. These provide an informed, but comprehensible, account of the major geological and geomorphological features. It would be good to see this series grow and reach wider audiences of interested visitors.

Several other activities have been designed to enhance both the scale and awareness of conservation in Namibia, and whilst these largely focus on bioconservation they also help focus attention of the importance of conserving geomorphological landscapes. For example, Namibia established Namibian Coast Conservation and Management (NACOMA) project in 2005. The objective of this was to strengthen conservation, encourage sustainable use of the coast's natural resources and to mainstream biodiversity conservation in coastal and marine ecosystems in Namibia. Funded by the Namibian Government and by Global Environment Facility (GEF), its aims were to:

- Enable Namibians to agree on a common vision for the management of the coastal zone
- Develop and support the implementation of the Government's coastal policy
- Clarify the legal and regulatory framework for coastal zone development planning
- Harmonize institutional mandates and roles for the management of the coastal zone
- Provide required training and practical skills to key stakeholders responsible for managing the coast
- Improve awareness about the coastal biodiversity, environmental problems and the coastal value

Similarly, the Namibia Protected Landscape Conservation Areas Initiative (NAM-PLACE), or 'Landscapes Namibia', is a five year project established by The Ministry of Environment and Tourism (MET), with co-financing from the Global Environment Facility (GEF) and with the United Nations Development Programme (UNDP) as the Implementing Agency. Started in November 2011, it aims to remove barriers to conservation, provide landscape scale conservation of biodiversity and build a more extensive protected area network. The project aims to establish new Protected Landscape Conservation Areas (PLCA) and also to formalize already existing ones by introducing collaborative governance structures. Each protected landscape in the scheme will link an existing State Protected Area with adjacent Communal Conservancies and Private Reserves. The project aims to add a further 15,550 km^2 of protected land beginning with five demonstration sites: Mudumu landscape, Greater Waterberg, the Windhoek Green Belt, the Greater Sossusvlei-Namib, and the Greater Fish River Canyon landscape. Funding and commitment from politicians and a wide range of stakeholders is vital if these sorts of schemes are going to be successful.

Engaging tourists with Namibian geomorphological landscapes is another vital task to increase awareness of their value and the threats to them, and thus contribute to conservation. In Namibia there are have been many recent attempts to increase the participation within the tourist industry of indigenous people and to develop their share of the economic benefits of tourism. Such community-based tourism activities can be a potent way of linking the different facets of conservation within Namibia—i.e. the conservation of biodiversity, geodiversity and cultural heritage. Not-for-profit groups such as Namibia Community Based Tourist Assistance Trust (NACOBTA) provide a forum for small scale community tourism initiatives, such as that at Spitzkoppe which provide a campsite and local tours around this major geosite.

References

Gray M (2013) Geodiversity, 2nd edn. Wiley-Blackwell, Chichester

Schneider GIC, Schneider MB (2004) Gondwanaland geopark. Proposal document. Accessed 21 Mar 2014 at http://portal.unesco.org/en/files/31729/11400823131Gondwana_Park_draft1corr.pdf/Gondwana+Park+draft1corr.pdf

Thomas MF (2012) A geomorphological approach to geodiversity-its applications to geoconservation and geotourism. Quaestiones Geogr 31:81–89

Wassenaar TD, Henschel JR, Pfaffenthaler MM, Nutota EN, Seely MK, Pallett J (2013) Ensuring the future of the Namib's biodiversity: ecological restoration as a key management response to a mining boom. J Arid Env 93:126–135

Index

A. Goudie and H. Viles, *Landscapes and Landforms of Namibia*, World Geomorphological Landscapes,
DOI 10.1007/978-94-017-8020-9, © Springer Science+Business Media Dordrecht 2015

MIX
Papier aus verantwortungsvollen Quellen
Paper from responsible sources
FSC® C105338

If you have any concerns about our products,
you can contact us on
ProductSafety@springernature.com

In case Publisher is established outside the EU,
the EU authorized representative is:
Springer Nature Customer Service Center GmbH
Europaplatz 3, 69115 Heidelberg, Germany

Printed by Libri Plureos GmbH
in Hamburg, Germany